INCIDENTAL TRAINER

A Reference Guide for Training Design, Development, and Delivery

INCIDENTAL TRAINER

A Reference Guide for Training Design, Development, and Delivery

Margaret Wan

CRC Press
Taylor & Francis Group
Boca Raton London New York

CRC Press is an imprint of the
Taylor & Francis Group, an **informa** business

CRC Press
Taylor & Francis Group
6000 Broken Sound Parkway NW, Suite 300
Boca Raton, FL 33487-2742

First issued in paperback 2019

© 2014 by Taylor & Francis Group, LLC
CRC Press is an imprint of Taylor & Francis Group, an Informa business

No claim to original U.S. Government works

ISBN-13: 978-1-4398-5790-8 (hbk)
ISBN-13: 978-0-367-37916-2 (pbk)

This book contains information obtained from authentic and highly regarded sources. Reasonable efforts have been made to publish reliable data and information, but the author and publisher cannot assume responsibility for the validity of all materials or the consequences of their use. The authors and publishers have attempted to trace the copyright holders of all material reproduced in this publication and apologize to copyright holders if permission to publish in this form has not been obtained. If any copyright material has not been acknowledged please write and let us know so we may rectify in any future reprint.

Except as permitted under U.S. Copyright Law, no part of this book may be reprinted, reproduced, transmitted, or utilized in any form by any electronic, mechanical, or other means, now known or hereafter invented, including photocopying, microfilming, and recording, or in any information storage or retrieval system, without written permission from the publishers.

For permission to photocopy or use material electronically from this work, please access www.copyright.com (http://www.copyright.com/) or contact the Copyright Clearance Center, Inc. (CCC), 222 Rosewood Drive, Danvers, MA 01923, 978-750-8400. CCC is a not-for-profit organization that provides licenses and registration for a variety of users. For organizations that have been granted a photocopy license by the CCC, a separate system of payment has been arranged.

Trademark Notice: Product or corporate names may be trademarks or registered trademarks, and are used only for identification and explanation without intent to infringe.

Visit the Taylor & Francis Web site at
http://www.taylorandfrancis.com

and the CRC Press Web site at
http://www.crcpress.com

Contents

Preface xv
Acknowledgments xvii
Biography xix

PART 1 Fundamentals of Training

Chapter 1 Introduction 3

    1.1   What's Wrong with That Training? 3
    1.2   Who Are "Incidental Trainers"? 4
    1.3   WIIFY—What's In It For You? 4
    1.4   What Are the Seven Keys to Successful Training? 5
    References 6

Chapter 2 Needs Assessment 7

    2.1   The Myth: Training Is the Panacea 7
    2.2   Needs Assessment 8
    2.3   Training Needs Analysis 8
        2.3.1   Goals of Training Needs Analysis 8
        2.3.2   Steps in Performing Training Needs Analysis 9
            2.3.2.1   Tour Facility 9
            2.3.2.2   Interview Personnel or Conduct Other Surveys 9
            2.3.2.3   Research Regulatory Requirements and Internal Policies 10
            2.3.2.4   Review Job Analysis and Training Record 10
            2.3.2.5   Determine Performance Gap and Desired Remedial Actions 11
            2.3.2.6   Characterize Training Audience 12
            2.3.2.7   Decide Training Topics to Be Covered 12
    2.4   Task Analysis 13
        2.4.1   Goals of Task Analysis 13
        2.4.2   Task Analysis Versus Job Description or Job Analysis 13
        2.4.3   Selection of Task Analysis Method 13
        2.4.4   Steps in Performing Procedural Task Analysis 14
            2.4.4.1   Identify Prerequisites from the Job Description 14
            2.4.4.2   Review Standard Operating Procedures 15

		2.4.4.3	Record Equipment Used and Read Manufacturer's Instructions............... 15
		2.4.4.4	Perform or Observe Someone Perform a Task... 15
		2.4.4.5	List All Steps in the Procedure................... 15
		2.4.4.6	Validate Task Inventory............................. 16
		2.4.4.7	Compare Actual Performance with the Standard... 16
	2.4.5	Intellectual Tasks... 16	
	2.4.6	Multiple Task Analyses 17	
2.5	Next Steps 17		
References 17			

Chapter 3 Anatomy of a Training Plan 19

 3.1 The Myth: No Planning Is Necessary 19
 3.2 A Basic Tool 19
 3.3 Structure of a Training Plan 19
 3.4 Alternative Formats and Advantages 21

Chapter 4 Learning or Performance Objectives 23

 4.1 The Cornerstone of a Training Plan 23
 4.2 Domains of Learning and Educational Objectives 24
  4.2.1 The Cognitive Domain—Bloom's Taxonomy 24
   4.2.1.1 Knowledge 24
   4.2.1.2 Comprehension 25
   4.2.1.3 Application 26
   4.2.1.4 Analysis 26
   4.2.1.5 Synthesis 26
   4.2.1.6 Evaluation 26
  4.2.2 The Affective Domain—Krathwohl's Taxonomy 26
   4.2.2.1 Receiving 27
   4.2.2.2 Responding 28
   4.2.2.3 Valuing 28
   4.2.2.4 Organizing 28
   4.2.2.5 Characterizing by a Value or Value Set 28
  4.2.3 The Psychomotor Domain 28
   4.2.3.1 Observing 29
   4.2.3.2 Imitating 30
   4.2.3.3 Practicing 30
   4.2.3.4 Adapting 30
 4.3 Meaningful Learning Objectives 30
  4.3.1 The ABCD Formula 31
   4.3.1.1 Audience 31
   4.3.1.2 Behavior 31

Contents

		4.3.1.3	Condition	31
		4.3.1.4	Degree	32
		4.3.1.5	Examples	32
	4.3.2	The SMART Principle		32
		4.3.2.1	Specific	33
		4.3.2.2	Measurable	33
		4.3.2.3	Actionable	33
		4.3.2.4	Results-Oriented	33
		4.3.2.5	Trainee-Centered	33
4.4	Next Steps			34
References				34

Chapter 5 Instructional Strategies ... 35

5.1	Instructional Strategies Defined		35
5.2	The Ubiquitous Lecture		35
5.3	The "Cone of Experience"		35
5.4	Learning Styles		36
5.5	Many Roads, One Destination		37
	5.5.1	On-the-Job Training	38
	5.5.2	Lecture and Panel	38
	5.5.3	Group Discussion	39
	5.5.4	Demonstration and Practice	40
	5.5.5	Role-Playing	40
	5.5.6	Self-Guided Discovery	41
	5.5.7	Collaborative Learning	41
5.6	Selection of the "Best" Strategies		42
	5.6.1	Learning Objectives	42
	5.6.2	Target Audience	42
	5.6.3	Trainer's Skills	43
	5.6.4	Situational Constraints	43
	5.6.5	Summary	43
References			44

Chapter 6 Training Aids and Media ... 47

6.1	The Double-Edged Sword		47
6.2	Unlimited Choices!		47
	6.2.1	Handout	47
	6.2.2	Slide Presentation	48
	6.2.3	Video/Audio	50
	6.2.4	Easel Pad, Dry Erase Board, Electronic Copyboard	51
	6.2.5	Model, Prop	52
	6.2.6	Costume	52
	6.2.7	Game	53
	6.2.8	Computer, Internet, Simulator	53

	6.3	Considerations for Choosing the "Best" Aids and Media 54
		6.3.1 Learning Objectives 55
		6.3.2 Target Audience 55
		6.3.3 Trainer's Skills 55
		6.3.4 Situational Constraints 56
		6.3.5 Summary 56
	References 57
Chapter 7	Physical Environment 59
	7.1	Turning (an Almost) Perfect Training Plan into a Bomb! 59
	7.2	Knowing the Basics of a Suitable Physical Environment 59
	7.3	Managing What You Can Control 60
		7.3.1 Advance Site Inspection 60
		7.3.2 Room Layout and Seating 61
		7.3.3 Lighting and Noise 62
		7.3.4 Climatic Conditions 62
		7.3.5 Water and Sanitation 63
		7.3.6 Safety and Evacuation Routes 63
		7.3.7 Equipment and Supplies 64
		7.3.8 Other Logistics 64
	7.4	Handling What You Cannot Control 65
		7.4.1 Distractions 65
		7.4.2 Furniture Design 65
	7.5	Pausing and Reflecting on Your Training Plan 66
	References 66
Chapter 8	Testing and Assessment 67
	8.1	Criticality of Testing and Assessment 67
	8.2	Approaches to Testing 67
		8.2.1 Pretest and Posttest 68
		8.2.2 Norm- and Criterion-Referenced Testing 68
		8.2.3 Formative and Summative Testing 69
	8.3	Reliability and Validity 70
	8.4	Testing Methods 71
		8.4.1 Multiple Choice, Multiple Select, True/False, Matching, or Ordering 72
		8.4.2 Fill-in-the-Blank or Short Answer 72
		8.4.3 Essay or Oral Explanation 72
		8.4.4 Case Study or Situational Judgment Testing 73
		8.4.5 Performance or Simulation 73
		8.4.6 Role-Playing 73
		8.4.7 Observation or Report 74
	8.5	Objectivity and Subjectivity 74
	8.6	Next Steps 74
	References 74

| Contents | ix |

Chapter 9 Presentation and Facilitation 77
    9.1 A Tale of Two Incidental Trainers.. 77
    9.2 Prior Proper Preparation.. 78
    9.3 Effective Presentation—You Should Be Nervous! 78
        9.3.1 Style.. 79
            9.3.1.1 Organization 79
            9.3.1.2 Vocabulary.. 80
            9.3.1.3 Vocal Variety.. 80
            9.3.1.4 Body Language... 81
            9.3.1.5 Visuals 82
        9.3.2 Purpose... 82
        9.3.3 Emotions... 83
        9.3.4 Audience... 83
        9.3.5 Knowledge.. 83
    9.4 Effective Facilitation—When Should You Stop Presenting? 83
        9.4.1 Asking Questions 84
        9.4.2 Responding to Questions.. 84
        9.4.3 Managing Behaviors 85
        9.4.4 Providing Feedback.. 85
    9.5 Future Improvement 86
    References 86

Chapter 10 Course Evaluation 87
    10.1 Why Ask for Criticism 87
    10.2 Who the Evaluators Are 87
    10.3 What Should Be Evaluated.. 88
        10.3.1 Trainer's Self-Evaluation.. 88
        10.3.2 Supervisor's or Auditor's Evaluation........................... 89
        10.3.3 Trainee's Evaluation 89
    10.4 How to Design a Course Evaluation Survey 90
        10.4.1 Question Design 90
            10.4.1.1 Closed-Ended and Open-Ended Questions.. 90
            10.4.1.2 Leading and Loaded Questions 91
            10.4.1.3 Word Usage.. 92
            10.4.1.4 Question Sequence....................................... 92
            10.4.1.5 Trainee/Evaluator Anonymity 92
        10.4.2 Response Rate 92
        10.4.3 Timing 93
    10.5 Which Data Are Relevant.. 93
    10.6 What Else Must Be Assessed 94
    References 94

Chapter 11 Program Validation and Continuous Quality Improvement 95
    11.1 An Integrated Training Program.. 95

11.2	Program Validation Purposes	96
11.3	Program Validation Criteria	96
	11.3.1 Level 1—Reactions	97
	11.3.2 Level 2—Learning	97
	11.3.3 Level 3—Behavior	97
	11.3.4 Level 4—Results	98
11.4	Program Validation Tools	98
	11.4.1 Course Evaluation and Test Result	99
	11.4.2 Reaction Survey	99
	11.4.3 Observation	100
	11.4.4 Controlled Experiment and Quasi-Experiment	100
11.5	Documentation	101
11.6	Continuous Quality Improvement (CQI)	102
11.7	Organizational Impact	103
References		103

PART 2 Training, Like the Pros

Chapter 12 Gaining Organizational Support 107

12.1	Understanding Value and Support	107
12.2	Identifying Training Costs	108
	12.2.1 Compensation	108
	12.2.2 Costs of Materials and Equipment	108
	12.2.3 Facility and Network Usage	108
	12.2.4 Other Expenses Associated with the Training Program	109
12.3	Recognizing Training Benefits	109
	12.3.1 Productivity	109
	12.3.2 Quality	109
	12.3.3 Safety	110
12.4	Calculating Cost–Benefit or Benefit–Cost Ratio	110
12.5	Analyzing ROI	111
	12.5.1 Single-Period ROI	111
	12.5.2 Multiple-Period ROI	111
	12.5.2.1 Payback Period	111
	12.5.2.2 Net Present Value (NPV)	112
	12.5.2.3 Internal Rate of Return (IRR)	113
12.6	Pinpointing the "Hot Button"	113
	12.6.1 Governing Board and Senior Management	113
	12.6.2 Middle Managers and Frontline Supervisors	114
	12.6.3 Employees	115
12.7	Presenting the Business Case	115
12.8	Summary	116
References		116

Contents　　　　　　　　　　　　　　　　　　　　　　　　　　　　　　　　　　　xi

Chapter 13　Testing with Validity and Reliability 117
　　　13.1　The "Weight of Evidence" 117
　　　13.2　Validity and Reliability 117
　　　13.3　Fixed-Choice Questions 118
　　　　　　13.3.1　Multiple Choice 119
　　　　　　13.3.2　True/False 120
　　　　　　13.3.3　Matching 120
　　　13.4　Open-Ended Questions 120
　　　　　　13.4.1　Fill-in-the-Blank and Short Answer 121
　　　　　　13.4.2　Essay or Oral Explanation 121
　　　13.5　Case Study and Situational Judgment 121
　　　13.6　Performance Assessment 122
　　　13.7　Cut Score 122
　　　13.8　Summary 123
　　　References 123

Chapter 14　Conducting Meaningful Surveys 125
　　　14.1　Why You Want to Know About Survey Design 125
　　　14.2　What Affects Data Quality 125
　　　　　　14.2.1　Survey Mode 126
　　　　　　14.2.2　Metric Validity 128
　　　　　　14.2.3　Question Content 128
　　　　　　14.2.4　Question Presentation 129
　　　　　　14.2.5　Response Rate 131
　　　　　　　　　　14.2.5.1　Survey Delivery 131
　　　　　　　　　　14.2.5.2　Survey Completion 132
　　　14.3　What to Check in a Pilot Test 133
　　　14.4　Summary 133
　　　References 133

Chapter 15　Leveraging Generational Learning 137
　　　15.1　Four Generations at Work 137
　　　　　　15.1.1　Silents 137
　　　　　　15.1.2　Baby Boomers 137
　　　　　　15.1.3　Generation Xers 138
　　　　　　15.1.4　Millennials 138
　　　15.2　Four Generations in Training 138
　　　　　　15.2.1　Training the Silents 138
　　　　　　15.2.2　Training the Baby Boomers 139
　　　　　　15.2.3　Training the Generation Xers 140
　　　　　　15.2.4　Training the Millennials 140
　　　　　　15.2.5　Training Four Generations Together 141
　　　　　　　　　　15.2.5.1　A Challenge and an Opportunity 141
　　　　　　　　　　15.2.5.2　An Example 142

Chapter 16 Training a Multicultural Work Force ... 145
- 16.1 Training in a "Flat" World ... 145
- 16.2 Understanding Cultural Diversity ... 145
  - 16.2.1 Conformity ... 146
  - 16.2.2 Gender Roles ... 146
  - 16.2.3 Uncertainty Acceptance ... 146
  - 16.2.4 Power Distance ... 146
- 16.3 Avoiding Cultural Pitfalls ... 147
  - 16.3.1 Respect the Trainees ... 147
  - 16.3.2 Apply Cultural Intelligence ... 148
  - 16.3.3 Speak and Write Simply ... 148
  - 16.3.4 Ensure Proper Translations ... 149
  - 16.3.5 Use Nonverbal Techniques with Discretion ... 149
  - 16.3.6 Employ Suitable Instructional and Communication Strategies ... 150
  - 16.3.7 Check for Comprehension ... 151
- 16.4 Summary ... 151
- Acknowledgment ... 151
- References ... 152

15.3 Summary ... 143
References ... 143

Chapter 17 Transitioning from Presenter to Facilitator ... 153
- 17.1 Focus on Trainee Achievement ... 153
- 17.2 Understand Learning Theories ... 154
  - 17.2.1 Information Processing and Cognitive Load Theory (CLT) ... 154
  - 17.2.2 Constructivism ... 155
- 17.3 Apply Instructional Design ... 157
- 17.4 Hone Communication Skills ... 157
  - 17.4.1 Two-Way Communication ... 157
  - 17.4.2 Active Listening ... 158
- 17.5 Practice Questioning Techniques ... 158
- 17.6 Handle Disruptive Trainees ... 159
- 17.7 Give Helpful Feedback ... 160
- 17.8 Summary ... 160
- References ... 161

Chapter 18 Achieving the Four E's of Training ... 163
- 18.1 The Four E's of Training ... 163
- 18.2 Educational and Entertaining Materials ... 163
  - 18.2.1 Icebreaker ... 163
  - 18.2.2 Multimedia Presentation ... 164

Contents   xiii

	18.2.3	Self-Guided Discovery	165
	18.2.4	Role-Playing	166
	18.2.5	Worked-Out Example	166
	18.2.6	Problem-Solving Activity	167
	18.2.7	Debate	168
	18.2.8	Game	168
18.3	Enthusiastic and Engaging Trainer		169
	18.3.1	General Discussion	170
	18.3.2	Breakout Session	170
18.4	Summary		171
References			171

Chapter 19 Directing an Energized Training Event 173

- 19.1 Get Ready 173
- 19.2 Promote Attendance 173
- 19.3 Handle Logistics 174
- 19.4 Arrive Early 174
- 19.5 Greet Trainees 175
- 19.6 Start on Time 175
- 19.7 Present the Program 176
  - 19.7.1 Events of Instruction 176
  - 19.7.2 Interactive Methods 176
  - 19.7.3 Visual Aids 177
    - 19.7.3.1 Slide Presentation 177
    - 19.7.3.2 Easel Pad 177
  - 19.7.4 Guest Speakers 178
- 19.8 End on Time 178
- 19.9 Summary 178
- References 178

Chapter 20 Adopting the New Paradigm: Virtual Training and M-learning 181

- 20.1 Definition of Virtual Training 181
- 20.2 Evolution of Distance Learning 181
- 20.3 Caveats in Implementing Virtual Training 182
  - 20.3.1 Suitability 182
    - 20.3.1.1 Advantages of Virtual Training 182
    - 20.3.1.2 Disadvantages of Virtual Training 182
  - 20.3.2 Trainee Assessment 183
  - 20.3.3 Course Design 184
- 20.4 Strategies for the Virtual Classroom 184
  - 20.4.1 Functionality 185
  - 20.4.2 Preparation 186
    - 20.4.2.1 Advance Site Inspection 186
    - 20.4.2.2 Room Layout and Seating 186

			20.4.2.3	Lighting, Noise, and Climatic Conditions 186
			20.4.2.4	Other Logistics 187
		20.4.3	Delivery 187
	20.5	Growth in M-learning 188
	20.6	Advantages of Mobile Apps 188
		20.6.1	Ease of Access 189
		20.6.2	Timeliness of Information 189
		20.6.3	Engagement of Trainees 189
		20.6.4	Support of Training Activities 189
		20.6.5	Chunking of Content 189
		20.6.6	Availability of Software 190
	20.7	Expectations of Mobile App Features 190
		20.7.1	Personalization 191
		20.7.2	Multimedia 191
		20.7.3	Interactivity 191
		20.7.4	Integration 191
		20.7.5	Support 192
	20.8	Design of Mobile Apps for Training 192
		20.8.1	Trainee Experience 192
		20.8.2	Screen Size 192
		20.8.3	Connection Speed 193
		20.8.4	Storage Capacity 193
		20.8.5	File Format 193
		20.8.6	Font Style 193
	20.9	Inclusion of Performance Support Tools 193
	20.10	Successful Deployment of Virtual Training 194
	20.11	Summary 194
	Acknowledgment 194
	References 195

Epilogue 197

Appendix A: Training Needs Analysis Sample Form 199

Appendix B: Task Analysis Sample Form 201

Appendix C: Training Plan Sample Form 203

Appendix D: Course Evaluation Sample Form 207

Index 209

Preface

Employee training occurs in every workplace. New employees require orientation. Experienced employees need additional skills. Then there is regulatory training mandated by law, such as safety training. Well-prepared training conducted by skilled instructors has a high success rate. Unfortunately, training is often carried out by subject-matter experts who have no knowledge or experience in the theories and practice of adult education. As a result, organizations find that their training programs are not producing the expected employee performance.

If you are a subject-matter expert that sometimes takes on the role of trainer, you are an "incidental trainer" and this book is written for you. Part 1 provides the fundamental steps in the design, development, and delivery of training. Part 2 discusses in greater detail some of the advanced training and facilitation techniques. In addition, please visit my personal website http://www.MargaretWan.com where sample forms, checklists, and other resources are available for download. The goal is for you, the incidental trainer, to be able to enhance your training and facilitation skills and run an effective training program just like a professional, full-time trainer. This success will in turn help to improve the job performance of your trainees and the productivity of your organization.

Acknowledgments

Some of the information in Part 1 of this book is adapted from *Fundamentals of Training: Design, Development, Delivery*, a workbook published by Better Trainers Inc. for its members. Appendixes A to D are reprinted with its permission. I am grateful to this nonprofit educational organization for allowing me to use the materials. In particular, John Morse, Cathy Naabe, and Tina White have been collectively and individually extremely supportive of this project and have devoted their time to review some of the contents of this work.

My sincere appreciation goes to Cindy Carelli and Jill Jurgensen. This project would not have come to fruition without their tremendous patience and valuable guidance. I also thank Jim McGovern, Michele Smith, and all the staff at CRC Press and Taylor & Francis who have worked diligently during the publishing process.

Margaret Wan

Biography

Margaret Wan, Ph.D., graduated from the University of South Florida in Tampa, Florida, where she received her master's degree and her doctoral degree in public health, with specialization in environmental and occupational health. She also holds a master's degree in health services administration from Nova Southeastern University in Fort Lauderdale, Florida, and a bachelor's degree in laws from the University of London, United Kingdom.

Along with her education in multiple disciplines, Margaret has diverse experience in different professions and industries. Her goal is always to help businesses improve productivity and profitability. Prior to entering the health and safety profession, she was management consultant at Acustar Consulting, assisting small businesses in profit enhancement. Her current position is Principal Consultant and Trainer, at EOH Consulting, an environmental and occupational health consulting firm. Her work includes employee training in health and safety issues as well as communication and leadership skills. She advocates and practices the four E's of training to maximize effectiveness: *entertaining* and *educational* materials plus *enthusiastic* and *engaging* trainer.

Margaret is a Certified Industrial Hygienist (CIH), Certified Hazardous Materials Manager (CHMM), Certified Healthcare Environmental Manager (HEM), and Certified Environmental, Safety and Health Trainer (CET). She has served as officer in several professional organizations including the chair of the Communication and Training Methods Committee and the president of the Florida Local Section of the American Industrial Hygiene Association, and program chair of the Training Technical Group of the Human Factors and Ergonomics Society. She is also an active member of the American Society of Safety Engineers and the National Environmental, Safety and Health Training Association. She is a frequent presenter and trainer at national and international conferences.

Margaret's "pastime" is her engagement in nonprofit educational organizations like Toastmasters International and Better Trainers, which help their members improve communication and leadership skills and training and facilitation techniques, respectively. She served on the board of directors of Toastmasters International from 2006 to 2008 and became the president of Better Trainers in 2010.

Part 1

Fundamentals of Training

1 Introduction

1.1 WHAT'S WRONG WITH THAT TRAINING?

The location was one of the many facilities of a large organization. It was time for a safety audit. Members of the audit team followed the protocol which, among other things, included reviewing compliance training documentation in various departments and then interviewing randomly selected employees to test their knowledge in health and safety policies and procedures. Such knowledge was especially important in a department where one of the auditors was visiting—the laboratory. Here, employees might be exposed to hazardous chemicals in their work. External government regulations and internal policies demanded that employees in the laboratory be familiar with potential chemical exposures and how to protect themselves from chemical spills and splashes.

The records looked good; just two weeks before, the safety manager of the facility conducted training of all employees of the laboratory. The auditor selected three employees and interviewed them separately, asking each of them questions from a structured list. Based on the training record, all the questions should have been covered in the training. This particular auditor was a new member of the audit team and was performing an audit for the first time. She expected that with the recent training, the employees would answer at least 80% of the audit questions correctly. To her surprise, the best score among the three employees was five questions correct out of a total of ten.

The auditor thought to herself, "Was there a mistake in the training record? Were these employees absent from the training two weeks ago?" That was unlikely, she decided. She knew the safety manager quite well and was convinced that he would maintain proper records meticulously. Another thought, "Perhaps the employees know I am new in the audit team and they are testing my patience ... But, that's probably not true either as they should know better. Their big boss, the chief operating officer, is serious about these audits and their results." She was puzzled. When the audit team reconvened in private, she mentioned this to her colleagues, whereupon they informed her that employees' low scores on the audit questionnaires were not uncommon. In fact, according to the audit team lead, safety managers frequently complained, "We have trained and trained and trained. [The employees] don't seem to get it!"

The auditor was me. The experience prompted me to carry out more observations and research the issue. As a result, I have come to the conclusion that employees' insufficient knowledge on topics they have been trained on is often a result of the trainer not knowing how to train. And, that is because the trainer is an "incidental trainer."

1.2 WHO ARE "INCIDENTAL TRAINERS"?

You may be one of them! If you are a professional in any field, at some point in your career or business you probably will be training employees or clients, making you an incidental trainer.

The term "incidental trainers" used in this book refers to professionals who have training as an incidental responsibility in their job or business. Most often they are not full-time trainers. They are subject-matter experts in diverse professions, such as engineering, law, safety, marketing, research, health care, risk management, information technology, and many others. Since they have the technical expertise, it is assumed that they can teach others on the subject matter. The safety manager at the audited facility mentioned above is one of them. His expertise is in health and safety in the workplace, not in training or adult education. Since he is the safety manager, however, safety training naturally falls in his lap. Other examples abound in almost every professional field and every organization. For instance, an experienced engineer may have the additional duty of providing on-the-job training to a new engineer, or a sales manager may need to train a team of sales representatives. Outside of the employment relationship, consultants and other business persons frequently assume the role of incidental trainers. A technology consultant may teach clients how to maximize the benefits from certain computer software; a communications expert may train his clients on how to use a newly installed telephone system. The training may be in a group setting or one-to-one.

1.3 WIIFY—WHAT'S IN IT FOR YOU?

In your responsibility as an incidental trainer, what happens if the training does not go well? It diminishes your credibility as a subject-matter expert. No matter how knowledgeable and experienced you are in the field, your trainees will not appreciate your expertise if they feel that they have not learned anything from your training. It could reflect poorly on your performance even though "training" may not have been specified in your job description!

How can you ensure the effectiveness of your training to satisfy your employers, clients, or trainees? Effective training:

Remediates deficiencies
Educates and entertains trainees
Stimulates higher performance
Uses skilled instructors
Lengthens retention time
Tests competencies
Supports organizational goals

In other words, effective training produces RESULTS.

This book will guide you through the process of training design, development, and delivery to help you achieve effectiveness in your training program. Since this book is written with the incidental trainer in mind, it emphasizes practical techniques that you can put to use easily. Incidental trainers typically spend less than half of their time in training and must manage that time wisely.

Part 1 of this book, "Fundamentals of Training," presents a step-by-step guide from training design to program validation. Part 2, "Training, Like the Pros," delves into specific topics and discusses best practices to help you enhance your training effectiveness to match those of professional trainers. At the same time, even full-time trainers may pick up new ideas that help them be more successful!

1.4 WHAT ARE THE SEVEN KEYS TO SUCCESSFUL TRAINING?

To be a successful trainer, you must first remember what training is all about. The Merriam-Webster Online (2011) defines "training" as "the act, process, or method of one that trains" and "the skill, knowledge, or experience acquired by one that trains." These definitions use the word "trains" from the perspectives of a trainer and a trainee. They imply that training is a process whereby the trainer imparts knowledge and skills to the trainee. Within the business world, another characteristic of training is that it must be supportive of an organization's goals. Hence, this book adopts the following definition:

> Training is a communication process involving interaction between trainer and trainee, the purpose of which is to transfer knowledge, teach skills, or transform behavior in support of the achievement of an organization's goals.

Note that the literature sometimes refers to changing "attitude" rather than "behavior." For example, Brethower (2000, 490) states that knowledge, skills, and attitudes are the "traditional domain of instructional content." Bloom's taxonomy (1956, 12), as explained in Chapter 4, defines learner behaviors to include ways in which learners think or feel. Also, feelings are not directly observable and must be measured by behavioral indicators. For these reasons, the term "behavior" is used in the above definition of training.

Is training different from teaching? Both training and teaching offer the learners an educational experience. As Smith and Ragan (2005, 5) state, training is distinguished from teaching in that the former focuses on skills that learners can immediately apply to improve job competency, whereas the latter targets skills of a more general nature that are not directed toward specific job tasks. Due to the similarity, instructional design techniques in teaching are applicable to training and vice versa. Due to the difference, one of the critical success factors in training is unique, and that is organizational support for the training program.

The seven keys to success of a training program are:

1. Organizational support
2. Needs assessment
3. Training plan
4. Competency assessment
5. Presentation and facilitation
6. Course evaluation
7. Program validation

These critical success factors apply whether you are providing training within your own organization or for your clients. By the same token, discussions in this book that make reference to "employee" are equally applicable whether the employees are your organization's employees or your clients or their employees.

While a validated training program helps to garner more organizational support, obtaining organizational support requires business management skills more than training skills. For this reason, the topic of organizational support is deferred to Part 2. The next chapter begins by looking at two types of needs assessment, both of which are necessary to determine if and what kind of training is desirable and how best to set the goals of the training.

REFERENCES

Bloom, B. S., ed. 1956. *Taxonomy of Educational Objectives: The Classification of Educational Goals*; *Book 1 Cognitive Domain*. New York: Longman.

Brethower, D. M. 2000. "The relevance of performance improvement to instructional design." In *The ASTD Handbook of Training Design and Delivery: A Comprehensive Guide to Creating and Delivering Training Programs—Instructor-Led, Computer-Based, or Self-Directed*, edited by G. M. Piskurich, P. Beckschi, and B. Hall. New York: McGraw-Hill, pp. 473–491.

Merriam-Webster Online, s.v. "Training," accessed July 26, 2011, http://www.merriam-webster.com/dictionary/training.

Smith, P. L., and T. J. Ragan. 2005. *Instructional Design*. 3rd ed. Hoboken: Wiley/Jossey-Bass Education.

2 Needs Assessment

2.1 THE MYTH: TRAINING IS THE PANACEA

At a staff meeting of an assembly plant, the quality assurance manager brought to the attention of the three foremen that the defect rate of products was creeping up in the past 8 weeks. His purpose was to solicit input as to what the solution might be. He asked the foremen, "How do you think we can correct this problem?" Without any further thought, all the foremen responded with the same suggestion, "We need to do more training to make sure the employees know how to do the job." The quality assurance manager did not stop there. He continued to question how long those employees had been on the job, what training they already received, and if anything had changed in the operations of the department. As he probed deeper, he discovered that the increased percentage of rejection was due to the machinery not having been properly serviced as per maintenance contract with an outside vendor. That vendor had gone out of business. Consequently, the machinery was not functioning as designed, yielding products of inferior quality. Obviously, it was not due to poor performance or lack of training on the part of the employees on the assembly line. Soon afterwards the plant manager identified a new vendor to take over the maintenance contract and the defect rate dropped back to within the acceptable range.

In another case in a different industry, the emergency department of a hospital had a recent adverse event, whereby an error in medication administration almost caused the death of a patient. In the corrective action plan that was part of the incident report, the charge nurse emphasized that training must be scheduled immediately to ensure that staff would be careful when picking up drugs from the supply cart and administering the medication to the patients. Upon reviewing the case and visiting the emergency department, however, the risk manager of the hospital pointed out that some of the packages of drugs looked so similar that in the hectic life of the emergency department, they would easily confuse even experienced, well-trained staff members. An interdisciplinary team including medical and risk management professionals redesigned the layout of the supplies on the cart, using color and shape coding to differentiate different drugs. Meanwhile, the procurement department tried to work with the drug suppliers to change the packaging as a long-term solution, since this facility was part of a large health care organization with a lot of bargaining power as a customer.

Oftentimes managers and supervisors have a tendency to consider training as a cure-all. Whenever something goes wrong, their immediate reaction is to give the employees more training, as if that will always correct the problem. There is no doubt that training is indispensable in any organization; however, it is not a panacea. As illustrated in the instances mentioned above, training, by itself, would not resolve the issues. In the first case, the machinery must be maintained to operate at peak performance. In the second case, redesign of the packaging using distinctly different shapes or colors for different drugs would be the ultimate solution. An organization can spend a lot of time and money in training and still does not achieve the desired outcomes if the lack of training was not the cause of the problem. A training needs assessment helps the organization avoid wasting resources by pinpointing areas where training will be truly beneficial.

2.2 NEEDS ASSESSMENT

A training needs assessment is "the process of collecting information about an expressed or implied organizational need that could be met by conducting training. The need can be a desire to improve current performance or to correct a deficiency" (Barbazette, 2006, 5). It is a critical success factor in effective training.

Many techniques are available for training needs assessment. For the incidental trainer who can only devote a percentage of his or her time to training, the two techniques that are most crucial are training needs analysis and task analysis. They target overall training needs for the training program and task-specific needs for particular training courses, respectively. Being subject-matter experts, incidental trainers are expected to analyze training needs within a department or functional unit and to carry out task-specific training, as opposed to overall employee training and development for the whole organization.

2.3 TRAINING NEEDS ANALYSIS

A training needs analysis is a thorough study of an organization, department, or functional unit to determine how training can help to improve effectiveness and efficiency and meet regulatory obligations or internal standards, thereby supporting the organization's goals and business needs. This definition differs from those most often found in published works on training; they normally mention the linkage between training and organizational goals but not departmental or regulatory requirements or internal standards. The rationale for the inclusion of these elements is the difference between target readers of the training publications and incidental trainers. The former are usually the staff in the training departments of organizations. By definition incidental trainers are not responsible for the training and development of the whole organization. They would probably focus on training within an operational department or functional unit that, in many cases, is directly responsible for implementation of compliance-related policies and procedures.

2.3.1 Goals of Training Needs Analysis

The goals of a training needs analysis are to find out:

- What employees know and what they should know
- How employees perform and how they should perform
- Why a discrepancy exists between the desired and actual results
- Whether and how training can correct that discrepancy

Needs Assessment 9

A training needs analysis enables you, the trainer, to verify that training will indeed contribute to overcoming whatever the problem is that your department or team is facing. You should ask the question: Is training the right solution to our organizational need, such as correcting deficiency or enhancing performance? As seen earlier, sometimes training may not be the true answer.

2.3.2 Steps in Performing Training Needs Analysis

Appendix A presents a training needs analysis sample form. The process of training needs analysis may involve one or more of the following steps depending on the nature of the job and the industry:

- Tour facility.
- Interview personnel or conduct other surveys.
- Research regulatory requirements and internal policies.
- Review job analysis and training record.
- Determine performance gap and desired remedial actions.
- Characterize training audience.
- Decide training topics to be covered.

2.3.2.1 Tour Facility

Studies have shown that the physical work environment has tremendous influence on productivity and creativity (Dul and Ceylan, 2011; Genaidy et al., 2009; Niemelä et al., 2002). Consequently, you may want to look at the work area and ask these questions:

- Are the workstation design, equipment, and supplies suitable for the job?
- Is there adequate and appropriate lighting?
- Does background noise cause distractions?
- Are temperature and humidity properly controlled?
- Are there any health or safety hazards in the work area?

For remote locations and with the increasing popularity of telecommuting, a site visit may not be cost-effective or feasible. If that is the case, you will have to rely on telephone interviews or other survey methodologies to collect the data.

2.3.2.2 Interview Personnel or Conduct Other Surveys

Surveys are important tools in the process of training needs analysis. Involve frontline employees, supervisors, and management to get their perspectives on aspects of the jobs, including the physical environment and job requirements. Managers can explain the expectations from management's perspective. Frontline employees and supervisors work in the field and are good sources of information. It is important to include frontline employees in the dialogue even if the target trainees are only supervisors or managers. Expectations of employees may influence how supervisors or managers react to an issue or rate employee performance (Shore et al., 1998). Be sure to include labor union representatives if applicable. These groups of stakeholders may hold very different opinions.

Before the survey, inform the employees why the needs analysis is necessary and how the data collected will be used to benefit them. This approach helps allay suspicion and improve learning motivation (Noe, 1986, 745). While conducting the training needs analysis, keep an open mind and remain as objective as possible.

Face-to-face or telephone interviews are the best in obtaining relevant responses. Practice good listening skills. Identify any environmental constraints. If interviews are not feasible, conduct a mail, e-mail, or web-based survey. Keep in mind that these survey responses are typically low and tend to suffer from bias more than personal interviews (Fowler et al., 2002). Many factors affect response rate and reliability in a survey (Anseel et al., 2010). For example, assuring potential respondents of anonymity may help increase the response rate (Faria and Dickinson, 1996). Web-based surveys present unique challenges (Fan and Yan, 2010). Chapter 14 discusses advantages and disadvantages of various survey methodologies and techniques to improve data quality.

In the course of the surveys, ask questions about the job tasks rather than training needs to discover the real challenge. A job task question for a sales assistant may be, "What steps do you take in designing a sales brochure?" In contrast, a training needs question may ask, "Have you attended any training class in graphic design?" Having attended training in graphic design does not guarantee that the employee has learned or can apply what has been taught in designing a brochure.

2.3.2.3 Research Regulatory Requirements and Internal Policies

Is there any discrepancy between regulatory requirements, internal policies, and what the employees are actually doing? Regulatory requirements applicable to the organization will take precedence over internal policies. In case of any conflict, the policies must be corrected. Assuming that there is no conflict, if deviations from policies are observed, the reason should be investigated. A valid reason justifies updating the policies, whereas an invalid reason demonstrates the need for training.

2.3.2.4 Review Job Analysis and Training Record

A job analysis is usually available from the human resource department. It specifies the knowledge, skills, abilities, and other characteristics (KSAOs) necessary to perform the job and the level of competence required. It helps you understand the job requirements and the disparity between those requirements and the job incumbent's KSAOs.

A previous training record shows what the employees were taught and how they were taught. If the employees were trained on the same topic and still lack the requisite KSAOs, find out why. It is possible that the earlier training strategies were ineffective, in which case this time you may have to devise different strategies. For instance, if the last training comprised lecture only, you may want to change the format to a combination of lecture and group activity to enhance audience participation and engagement. Previous course evaluations by trainees would be helpful as well, if available.

Needs Assessment

2.3.2.5 Determine Performance Gap and Desired Remedial Actions

Analyzing data collected in the previous steps should reveal the root causes of the gap between desired and actual performance. Suitable remedial actions may or may not include training—the purpose of this whole exercise is to find out if the need exists.

Consider this scenario: A company manufactures and installs custom industrial shelving systems. The senior sales engineer has noticed that the sales in the last 3 months have not met the budgeted growth. In fact, they have fallen slightly behind the sales for the same period last year. Four of the ten sales engineers have been with the company for less than a year. Have the sales dropped because the senior sales engineer has not trained these new sales engineers sufficiently on how to sell the products?

Perhaps, but the senior sales engineer should not rush into planning a 3-day intensive sales training for his team until he confirms that lack of sales training is the cause of the disappointing results. A better approach is to conduct a training needs analysis first. He schedules a series of telephone meetings with his sales engineers who are located in different regions of the country. During these meetings, he tries to verify what each of them knows about selling and about the products. He solicits input from the team members as to what issues they may be facing in closing sales. As it turns out, the biggest hurdle is that the production department has changed the finish paint on the shelving system, giving it a different look and texture. Customers perceive the new look as equivalent to inferior quality and are switching to competitors' products. In this situation, sales training may not improve sales performance, except that encouraging proactive and timely feedback from team members that interact with customers may be helpful.

In a slightly different scenario, the product has not changed in the past year. However, the senior sales engineer discovers that the new team members have not gained enough product knowledge to talk to potential buyers and have been just taking orders from existing customers. They lost a few major customers when these customers decided to close their plants. Not enough new customers have been established to fill the void. In this case, training the sales force with the goal of improving product knowledge and prospecting skills will be useful.

In general, training is not the solution to performance issues when organizational factors affect employee morale or impose environmental constraints, or when nonwork-related issues come into play. If an organization is going through rightsizing, the remaining employees may be physically and mentally stressed due to additional workload and concern over further layoffs. Low morale caused by these factors leads to poor performance and cannot be corrected with training. Environmental constraints are workplace factors that are beyond the control of the individuals working in that area. The machinery maintenance, drug packaging, and finish paint in the scenarios previously described are examples. Nonwork-related issues could be personal or family problems that affect decision making at work. Unfortunately, they are occurring more and more often for reasons such as the prevalence of obesity and related employee health concerns, or employees in the "sandwich generation" serving as caregivers of younger and older family members.

Training may be the solution in improving performance when there is a gap between what employees know and what they should know, or between how they perform and how they should perform, after taking into consideration organizational or personal factors. In these situations transferring knowledge or skill or changing behavior can improve productivity and profitability. An employee who is reassigned to a new task should be trained. Employees should also be trained when new equipment, process, or procedure is introduced. Training may be required by law, such as sexual harassment or safety training, or it may be dictated by company policy, as when a company requires all new employees to attend orientation. In summary, training is appropriate for:

- Performance gap correctable by knowledge/skill transfer or behavior change
- Regulatory compliance
- Internal policy requirement

All stakeholders concerned—managers, supervisors, employees, or their unions—should agree on the remedial actions. They should understand what training will do and what it will not do. If the consensus is that training will be beneficial, the next steps will help you plan an effective training program.

2.3.2.6 Characterize Training Audience

Identify the potential trainees by asking these questions:

- Who should be trained?
- What are their job titles and functions?
- Do they work shifts? What is the best time to do the training?
- Are they new or existing employees?
- What are their education level, gender, and age range?
- Does the group include employees who do not understand English very well?
- Does anyone need special accommodation due to disabilities?

Only the job title is recorded on the training needs analysis sample form in Appendix A to keep it simple. Other characteristics are noted on the training plan sample form in Appendix C that is used for each training course.

2.3.2.7 Decide Training Topics to Be Covered

The final step in any training needs analysis is to decide what topics should be covered in the training. It must be emphasized that the training needs analysis is a macro analysis; therefore, the training topics decided at this stage are broad. For example, you and other stakeholders may agree that certain employees need training in computer skills to improve proficiency and cope with the workflow in the department, but it has not been determined exactly what they must learn and how they will learn it. That will be resolved after a task analysis.

Needs Assessment

2.4 TASK ANALYSIS

A task analysis is a process whereby you examine a work activity in order to define what you want your trainees to learn.

2.4.1 Goals of Task Analysis

The goals of a task analysis are to determine:

- What trainees should learn
- What the criteria of successful learning are

A skilled trainer must identify the learning outcome—how the trainees should be able to think and perform—and this cannot be achieved without a task analysis (Jonassen et al., 1998, 1). A task analysis is the foundation for developing your training plan. It provides you with information that forms the basis of the instructional design—the learning objectives, instructional strategies, training media, training environment, and competency assessment (Gagné, 1962, 88). A task analysis is also vital to ensure that the correct steps of a task will be taught in the correct sequence during the training.

2.4.2 Task Analysis Versus Job Description or Job Analysis

Why should you perform a task analysis? Why not use the job description or job analysis that is readily available from human resource records? A job description or job analysis compiled for human resource management is typically used for job classification, compensation, and evaluation. It stipulates the content and requirements of the job, identifying the tasks, duties, and responsibilities, but does not have enough details on how they are performed. Although job analysis is actually one method of task analysis, unless a job analysis is conducted with instructional design in mind, it would not have the kind of detailed information that you need to form the basis of training design. For example, it will not state the sequential steps in performing a task. That is not to underrate the role of job analysis as an essential process for planning employees' training and development (Prien et al., 2009, 9). As stated in the steps of training needs analysis, a job analysis provides valuable information on the KSAOs. It is just that the training responsibilities of incidental trainers are focused on the operational aspects of the job, whether it is in sales or finance or any other profession, and a more specific analysis of the work activity—the task—within the context of the job is required.

2.4.3 Selection of Task Analysis Method

Many methods of task analysis are available and used by various disciplines for various purposes, such as training, system design, and safety evaluation. Terminologies in the literature also differ and overlap. For the purpose of instructional design, selection of the best method depends on the training goals and the nature of the job. An incidental trainer does not have the time or resources to research or apply complex methods. One method that can be employed in many job situations is introduced here in a simple, easy-to-use format. It is a technique derived from procedural task analysis.

2.4.4 Steps in Performing Procedural Task Analysis

A work activity has a definite beginning and ending and leads to a product, service, or decision. A procedural task analysis aims at compiling a task inventory, which is a sequential listing of all steps necessary to perform a work activity successfully. The analysis also identifies any prerequisite skills, supplies, and equipment that are necessary to perform the task, as well as environmental constraints that may hinder proper performance.

Appendix B presents a task analysis sample form. It shows that a task has a goal and three phases. The goal is what the activity tries to accomplish. The three phases are preparation, performance, and follow-up. For example, a technician has to repair a lathe. The goal is to make that lathe work again. The preparation phase could be troubleshooting and assembling the supplies and tools required for the repair. The performance phase is the actual repair. The follow-up phase could be cleaning up and returning the tools to storage. All steps taken during the preparation, performance, and follow-up phases should be included in the task inventory. Additionally, the sample form documents the review of relevant documents, prerequisite skills, supplies, and equipment needed to properly perform the task, and any environmental constraints.

A procedural task analysis may include one or more of these steps:

- Identify prerequisites from job description.
- Review standard operating procedures.
- Record equipment used and read manufacturer's instructions.
- Perform or observe someone perform task.
- List all steps in procedure.
- Validate task inventory.
- Compare actual performance with standard.

These steps are explained in the following sections.

2.4.4.1 Identify Prerequisites from the Job Description

Although the availability of a job description does not negate the need for a task analysis, a job description can provide you with information on prerequisites such as KSAOs and supplies and tools used. If a job description is not available within the organization, the database of the Occupational Information Network (O*NET) of the U.S. Department of Labor has job descriptions of many occupations. For example, according to O*NET (U.S. Department of Labor, 2010), an industrial designer, among other things, should have knowledge of design techniques and principles in the production of technical plans, skills in active listening and critical thinking, and abilities in oral comprehension and deductive reasoning; tools used in the job include a computer and digital camera. You should study the job description when you perform a task analysis and note the most important prerequisites.

2.4.4.2 Review Standard Operating Procedures

Standard operating procedures, if available, can shed light on how the task should be performed. Be cognizant of the fact that sometimes standard operating procedures are not updated as often as they should be or they may not meet regulatory requirement. They do provide a starting point to help you understand the steps in a task.

2.4.4.3 Record Equipment Used and Read Manufacturer's Instructions

If the task involves operating certain equipment, review the manufacturer's instructions or user guide. In almost all cases, employees should be taught to perform tasks according to those instructions; otherwise, warranties may be invalidated or, even worse, property damage or personal injury may result. If employees or supervisors have modified the procedures specified in a manual, the manufacturer should be consulted first before teaching operators of the equipment to do things differently.

2.4.4.4 Perform or Observe Someone Perform a Task

This step may seem obvious in a task analysis, but occasionally an incidental trainer makes a mistake of developing the training based on the standard operating procedures for a task, not realizing that the actual practice on the job has changed for a long time and for a good reason, such as process change. It is essential always to perform or observe the task being performed to verify the steps actually used in day-to-day operations. If you are conversant with the task, you can perform it. If not, watch an experienced job incumbent performing the task.

When observing a manual task, you may find it helpful to make a video recording of the task performance during the analysis. Video recording has been used in ergonomics research and time-and-motion studies because of the convenience of replaying the video to examine the exact movements and lapsed time. If video recording is done, ensure that legal ramifications are considered and requisite releases are obtained from people and owners of property that are identifiable in the video.

2.4.4.5 List All Steps in the Procedure

It is important that the task inventory lists all steps in the procedure. Special attention is appropriate in an on-the-job training situation where an incidental trainer routinely performs the task in question. The trainer may be so familiar with the task that recording of some steps is overlooked because they have become second nature. If you are in such a situation, try to reflect on every step of the task and think about questions someone unfamiliar with the task may ask.

Use action verbs and short sentences to describe the steps. An example may be "Access hazardous materials database on computer." Include all steps in the preparation, performance, and follow-up phases.

Sometimes the task is broken down into subtasks that reflect subgoals, as done in a hierarchical task analysis (Shepherd, 2001, 1). Whether subtasks should be listed in a task analysis depends on the complexity of the task and the knowledge level of the trainees. In the previous example of accessing a database, if trainees may be unfamiliar with using the specific software application, subtasks should be listed, such as "Double-click HAZMAT icon on the screen to open program," "Log in using network user name and password," and so on. In contrast, if trainees who are familiar with the sign-on procedure are to be trained on performing specific searches within the database, such detailed steps are unnecessary.

During the observation, note any deviation from standard operating procedures or manufacturer's instructions.

2.4.4.6 Validate Task Inventory

No matter who has performed the task during the task analysis, the resulting task inventory should be verified by "a second pair of eyes" to ensure accuracy. A job incumbent other than the person who performed the task during the analysis would be an ideal person to verify the list. If no one else has the same job or performs the same task, have another person view the task performance or video recording and compare it with the list. In certain industries or jobs, validation may be necessary for safety or liability reasons.

2.4.4.7 Compare Actual Performance with the Standard

Any deviations of the actual performance from the standard should be noted in the task analysis. Deviations may occur in many dimensions, such as timeliness, quality, and quantity. You may want to obtain feedback from the stakeholders before finalizing the task inventory. Once that has been done, your objective will be to correct unwarranted deficiencies in the upcoming training.

2.4.5 INTELLECTUAL TASKS

You may ask, "What if the task is not a manual task? Mental steps cannot be observed." A task analysis can and should be done for manual and intellectual tasks since both have the same elements including the three phases of preparation, performance, and follow-up. It is acceptable for you to record the mental steps and decision process by relying on one or more persons performing the task to describe the steps.

For example, in the task of submitting an accident report, the goal is to record the accident. The preparation phase could be identifying the details that must be included and deciding who should receive the report. The performance phase is the process of writing and distributing the report. Follow-up action could be obtaining acknowledgment of receipt of the report from the recipients. Another example is the task of coaching an employee, with the goal of eliminating tardiness. The preparation phase could be retrieving an attendance record. The performance phase could include these steps: meeting with the employee, pointing out the discrepancy between expectations and performance, asking open-ended questions, listening attentively, and providing constructive feedback. The follow-up phase could be tracking future attendance. In these examples, as in most tasks, the activities have both physical and mental components. In all cases, the task analysis should define the goal of the task and break down the task into the constituent steps so as to determine what the trainee needs to learn to perform the task properly.

Special techniques in cognitive task analysis are applicable to tasks that require reasoning, organizing information, and making decisions (Crandall et al., 2006, 3). They are often used in the design of virtual reality training.

2.4.6 Multiple Task Analyses

If the training needs analysis conducted earlier has revealed that multiple training topics are needed, it may be necessary to analyze more than one task. For instance, a laboratory has a recent incident of workplace injury due to chemical burns. Sulfuric acid, a highly corrosive substance, was splashed onto an employee's hands while he was pouring the acid into a sink. The laboratory manager carried out a training needs analysis along with the accident investigation. She noted two problems that could be corrected by employee training. First, the gallon container of sulfuric acid was not properly labeled, and the injured employee thought that he was only pouring water down the drain. Second, instead of holding the container carefully and stabilizing it on the counter top adjacent to the sink while pouring, the employee poured the liquid from mid-air about a foot above the top edge of the sink. If it were not for the personal protective equipment he was wearing, the injury would have been much more severe. A frontline supervisor reported that some employees were not aware of standard operating procedures because they were not performing their regular jobs—they were substituting for other employees who were on vacation. The laboratory manager consulted with the supervisors. As a group, they determined that due to a staffing shortage, it was impossible to avoid having substitute employees (an environmental constraint), and the only way to ensure their competency and safety would be to cross-train them. The training topics would include the proper labeling of chemicals and handling of liquids, two distinct job tasks requiring separate task analyses.

2.5 NEXT STEPS

Having confirmed the training need and completed the task analysis, you are ready to begin work on the training plan. The next chapter provides an overview of the elements of a training plan.

REFERENCES

Anseel, F., F. Lievens, E. Schollaert, and B. Choragwicka. 2010. "Response rates in organizational science, 1995–2008: A meta-analytic review and guidelines for survey researchers." *Journal of Business and Psychology* 25 (3): 335–349.
Barbazette, J. 2006. *Training Needs Assessment: Methods, Tools, and Techniques*. San Francisco: Pfeiffer.
Crandall, B., G. Klein, and R. R. Hoffman. 2006. *Working Minds: A Practitioner's Guide to Cognitive Task Analysis*. Cambridge: The MIT Press.
Dul, J., and C. Ceylan. 2011. "Work environments for employee creativity." *Ergonomics* 54 (1): 12–20.
Fan, W., and Z. Yan. 2010. "Factors affecting response rates of the web survey: A systematic review." *Computers in Human Behavior* 26 (2): 132–139.

Faria, A. J., and J. R. Dickinson. 1996. "The effect of reassured anonymity and sponsor on mail survey response rate and speed with a business population." *Journal of Business and Industrial Marketing* 11 (1): 66–76.

Fowler, F. J., Jr., P. M. Gallagher, V. L. Stringfellow, A. M. Zaslavsky, J. W. Thompson, and P. D. Cleary. 2002. "Using telephone interviews to reduce nonresponse bias to mail surveys of health plan members." *Medical Care* 40 (3): 190–200.

Gagné, R. M. 1962. "Military training and principles of learning." *American Psychologist* 17 (2): 83–91. doi:10.1037/h0048613.

Genaidy, A. M., R. Sequeira, M. M. Rinder, and A. D. A-Rehim. 2009. "Determinants of business sustainability: An ergonomics perspective." *Ergonomics* 52 (3): 273–301.

Jonassen, D. H., M. Tessmer, and W. H. Hannum. 1998. *Task Analysis Methods for Instructional Design*. Mahwah: Lawrence Erlbaum.

Niemelä, R., S. Rautio, M. Hannula, and K. Reijula. 2002. "Work environment effects on labor productivity: An intervention study in a storage building." *American Journal of Industrial Medicine* 42 (4): 328–335.

Noe, R. A. 1986. "Trainees' attributes and attitudes: Neglected influences on training effectiveness." *Academy of Management Review* 11 (4): 736–749.

Prien, E. P., L. D. Goodstein, J. Goodstein, and L. G. Gamble, Jr. 2009. *A Practical Guide to Job Analysis*. San Francisco: Pfeiffer.

Shepherd, A. 2001. *Hierarchical Task Analysis*. New York: Taylor & Francis.

Shore, T. H., J. S. Adams, and A. Tashchian. 1998. "Effects of self-appraisal information, appraisal purpose, and feedback target on performance appraisal ratings." *Journal of Business and Psychology* 12 (3): 283–298.

United States Department of Labor. 2010. "Summary Report for: 27-1021.00—Commercial and Industrial Designers." Accessed July 30. http://www.onetonline.org/link/summary/27-1021.00

3 Anatomy of a Training Plan

3.1 THE MYTH: NO PLANNING IS NECESSARY

Will an architect start a construction project without any blueprints? Will the navigator of a cruise ship take the crew and passengers on a journey without charting the course? Yet the same professionals, when given the role of incidental trainers, have been known to conduct training without planning. The individual figures that he or she knows architecture or navigation so well that teaching anyone about the subject at any time is not a problem. No preparation is needed!

That is another myth about training.

3.2 A BASIC TOOL

A training plan is like the blueprints in a construction project or the navigation charts for a voyage. It is a basic tool used to ensure that the training goals will be met. Unexpected events do occur in projects. Flexibility to adapt to changes during execution of a plan is a desirable skill. On the other hand, flexibility is not equivalent to lack of planning. A navigator may have to change the course of a ship temporarily because of an approaching hurricane. The ultimate destination does not change. With the navigation charts, the ship can be directed back on course heading toward the destination once the storm is over. Similarly, you may need to modify planned activities due to the audience's reaction while you are delivering the training. A training plan will help you maintain focus on the training goal even though the means to the end may have changed.

A training plan can be developed for an organization or department as a whole or for a specific training course. At the organizational or departmental level, a training plan guides the entity in its overall strategies for employee growth and development. For each training course within a training program, a training plan enables you to focus on what needs to be covered in the course and how to achieve that goal. Incidental trainers are likely to need the second type of training plan; therefore, it will be the basis for the discussions of training plans in this book.

3.3 STRUCTURE OF A TRAINING PLAN

Appendix C presents a sample training plan. Although the sample form illustrates a training plan that is presented in a linear fashion, many parts of it are interrelated and interdependent, just like the parts in the structural makeup of an organism. If the demographics of the target audience change, the instructional strategies may change. Or, when the preferred physical environment cannot be used, you may need a different training aid. This may happen, for instance, when training is moved from an outdoor to an indoor environment due to seasonal variation in climatic conditions, and a replica of a large piece of outdoor equipment must be used inside a building instead of the real equipment installed outside.

The following are the major parts of a training plan:

1. General information—this part shows the title of the course, with an indication whether it is new or revised, the frequency of providing this training, and what the course is about. If the course is related to a specific task, it is a good idea to attach the task analysis to the training plan for easy reference. The agenda for the training event can be included as well when it becomes available. As explained in Section 3.4, having an agenda has advantages.
2. Target audience—the timeless edict, "Know your audience," is as vital for trainers as for speakers. The information in this section is derived from the needs assessment that has been completed. Listed are the job titles or functions and the related prerequisites as well as the knowledge, skills, abilities, and other characteristics (KSAOs) required, along with an indication whether the individuals work shifts. In some organizations, it is more important to consider the job functions than the job titles. In theory job titles should describe job functions performed. In practice job titles are often used to classify jobs for pay purpose. A "research analyst" may be someone collecting and analyzing financial market data, or performing a population health survey, or formulating models in operations research. These are quite distinct job functions. Some topics that have broad applications may be taught to a transdisciplinary group. Examples are office ergonomics, business computing, and time management. If that is the case, make a note of the diverse audience in the training plan. Other known attributes of the trainees should be included. One attribute that is helpful, if known, is the trainees' motivation to attend the course. Are they eager to learn more about the topic, or do they attend training only because required by their supervisors or company policies? Did they have a disappointing experience with previous training attended? Try to gain some insight into the demographics of the audience. Such information may be easier to obtain if the training is specific to one job than if the trainees come from a variety of departments or organizations. Look for answers to these questions:
    - What education background do the trainees have?
    - What is the audience's gender distribution?
    - What are the age groups?
    - Are there persons who do not speak English (or whatever language is used in the training)?
    - Does any individual need special accommodation to attend the training?
3. Learning or performance objectives—the objectives are probably the most important part of the plan and will be addressed in Chapter 4.

Anatomy of a Training Plan

4. Successful course completion criteria—what must the trainees achieve to be considered having passed the course or reached the level of competency expected of them? The criteria go along with the learning objectives and will be discussed in Chapter 4.
5. Continuing education credit—if the training qualifies for continuing education credit for the trainees' professions, plan to issue certificates of satisfactory completion that indicate the hours or points.
6. Instructional strategies—how to choose instructional strategies will be the topic in Chapter 5.
7. Training aids, media, and other equipment—these items are indispensable in training and will be reviewed in Chapter 6.
8. Physical environment—selecting the right setting and managing the physical environment will be the focus of Chapter 7. Virtual training will be discussed in Chapter 20.
9. Testing methods—testing techniques will be examined in Chapter 8.
10. Course evaluation and trainer's observations—these tools for continuous improvement will be explained in Chapter 10.
11. Other comments—any other important considerations should be noted in this space.

3.4 ALTERNATIVE FORMATS AND ADVANTAGES

Instead of the two-page training plan as presented in the sample, a more detailed training plan can be prepared to serve as an outline of how the course will be taught and what materials will be covered. Alternatively, an agenda can be appended to the training plan. In addition to its primary purpose of ensuring training goals are met, a training plan with an outline or agenda also helps you:

- Present the course contents in a logical order
- Avoid omitting essential portions inadvertently
- Emphasize materials relative to their importance
- Allot proper time for each instructional strategy
- Provide for trainee participation
- Run the training event on schedule

None of these matters, however, if the learning objectives in the training plan are erroneous. The next chapter examines why and how learning objectives should be established.

4 Learning or Performance Objectives

4.1 THE CORNERSTONE OF A TRAINING PLAN

An oft-cited quotation by former U.S. Secretary of State Dr. Henry Kissinger says, "If you don't know where you are going, every road will get you nowhere." In everything a person does professionally and personally, the person must know what the aims are in order to achieve them. Training is no different. Trainers must articulate the objectives of the training so that they are clear to the trainers themselves and to the trainees. The objectives may be framed as learning objectives if they relate to transfer of knowledge or, when the expected outcome of the training is improved job skills, they are often framed as performance objectives. In every case they should match the training needs of the audience.

Sometimes a distinction is made between "learning objectives" and "learning outcomes." For the incidental trainer, there is no practical difference as the purpose of the training is to fill the gap between what trainees know or do and what they should know or do, with the ultimate goal of enhancing performance or correcting deficiency. The terms are used interchangeably in this book, although it is acknowledged that there are circumstances where a time lag exists between completion of the training and measurement of the learning outcome.

For the trainer, having clear, focused learning objectives goes a long way to facilitate completion of the rest of the training plan and avoid wasting time on irrelevant or unimportant topics during the development or delivery of the training. The objectives form the basis of measuring whether the purpose of the training is achieved.

For the trainees, the learning objectives inform them what they expect to learn and, if introduced effectively prior to or at the beginning of the training event, create anticipation and excitement that will enhance the learning experience.

Writing learning objectives is one of the most challenging tasks for many incidental trainers. That is because most incidental trainers lack a good understanding of the domains of learning. Additionally, incidental trainers tend to write learning objectives from the perspective of what they want to accomplish during the training event, rather than what their trainees should achieve. To properly establish the learning objectives of a training course, an incidental trainer should have some basic knowledge of the domains of learning and the taxonomy of educational objectives. The next section offers a brief description.

4.2 DOMAINS OF LEARNING AND EDUCATIONAL OBJECTIVES

Between 1949 and 1953, a committee of college and university examiners collaborated to build a taxonomy of educational objectives in a similar fashion to biological taxonomies that classify animals and plants. The intent was to assist teachers, administrators, and researchers in defining educational outcomes and to facilitate information exchange among these parties on issues of curriculum and evaluation development (Bloom, 1956, 10). The project and later research by other authors resulted in several classic references on the cognitive, affective, and psychomotor domains of educational objectives. These three domains with regard to a task are not mutually exclusive. A particular task may encompass educational objectives across two or more domains. Some taxonomies also added a fourth domain, the social domain, treating it as distinct and separate from the cognitive and affective domains (Dettmer, 2006).

According to Bloom (1956, 12), the taxonomies classify *intended* learner behaviors—"the ways in which individuals are to act, think, or feel as the result of participating in some unit of instruction." Actual behaviors may differ from the intended behavior, reflecting failure to achieve the learning objectives.

4.2.1 THE COGNITIVE DOMAIN—BLOOM'S TAXONOMY

After more than 50 years, what is commonly referred to as Bloom's (1956) taxonomy of cognitive domain is still widely used in education and training. In recent years, a revised taxonomy has been published (Krathwohl, 2002), changing the organization of and some terminology in this classification, but the concepts remain the same.

The cognitive domain is concerned with remembering, thinking, and reasoning. Figure 4.1 illustrates the six levels of learning outcomes within the cognitive domain, along with examples of action verbs that may be used in stating learning outcomes at each level. The levels reflect learning outcomes that begin with simple behaviors at the lowest level to complex behaviors at the higher levels (Bloom, 1956, 18). In other words, each higher level requires skills and abilities acquired at the levels below. However, some action verbs may be suitable for more than one level depending on the context.

4.2.1.1 Knowledge

Knowledge is the lowest level of learning outcomes. It emphasizes "remembering, either by recognition or recall, of ideas, material, or phenomena" (Bloom, 1956, 62). The individual is expected to recall specific or general information. Specific information may be names or processes. General information may include patterns or generalizations that are used to solve a problem, although the individual is not expected to be able to apply the theories of problem solving. In a work situation, for example, an employee in a customer service team may need to know the telephone number to call a contractor to pick up secured storage bins of confidential papers for shredding. The employee may know that this contractor has been selected based on integrity, efficiency, and location. The employee does not need to apply these

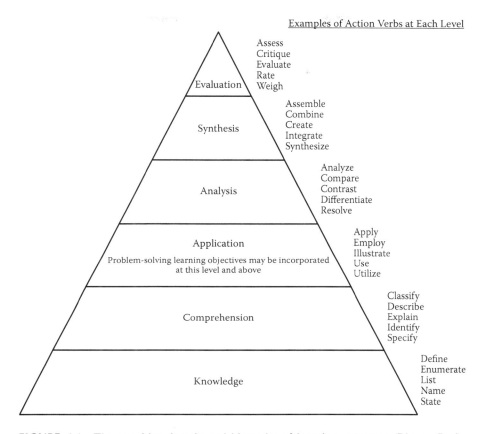

FIGURE 4.1 The cognitive domain and hierarchy of learning outcomes (Bloom, B. S., ed. 1956. *Taxonomy of Educational Objectives: The Classification of Educational Goals; Book 1 Cognitive Domain.* New York: Longman). Note that an action verb may be suitable for more than one level of learning outcome depending on the context.

criteria to choose from potential contractors as that would be the responsibility of the procurement staff.

4.2.1.2 Comprehension

Comprehension is the next level in the hierarchy of learning outcomes in the cognitive domain. The individual is expected to know the ideas communicated and to use them, but is not expected to be able to relate those ideas to other materials or understand the full implications (Bloom, 1956, 89). To gain understanding at this level, the individual may translate, interpret, or extrapolate the materials. These thought processes allow the individual to accurately paraphrase the materials in the original communication, understand the meaning of metaphors and symbolism, or translate words to symbols and vice versa. For instance, an employee at a warehouse handling export of merchandise should be trained to understand what the international shipping symbols mean.

4.2.1.3 Application

The step-up from comprehension to application is demonstrated by the fact that instead of being able to correctly use a theory, idea, or method only when it is the specified solution, the individual can determine when to use the abstraction to solve a problem in a suitable situation (Bloom, 1956, 120–121). The application level of learning outcomes is important in training because most training is intended to teach trainees how to use certain skills in the real world. A nursing assistant who has been trained on patient assessment and various types of safe patient-handling equipment will have achieved the desired learning outcome at the application level if he or she is able to select and use the proper piece of equipment to transfer a patient after assessing the patient's mobility status and medical condition.

4.2.1.4 Analysis

Analysis may be considered "as an aid to fuller comprehension or as a prelude to an evaluation of the material" (Bloom, 1956, 144). At this level of learning outcomes, a learner is able to analyze individual elements, the relationships between the elements, or organizational principles among those elements (145). The learner does this by breaking down the materials into parts to understand the structure. For example, a marketing manager may train an assistant to analyze competitors' marketing activities and presence in the community in relation to their sales and market shares.

4.2.1.5 Synthesis

Synthesis is the combining of elements or parts to form a unique communication, such as a speech or a white paper, to produce a plan that satisfies specified objectives, or to derive a set of abstract relations. Comprehension, application, and analysis also involve the combination of elements and construction of meanings, but the magnitude of the task is not as complete as in synthesis (Bloom, 1956, 162). Synthesis produces something that is new and emphasizes originality and uniqueness. For example, in a train-the-trainer program, the development of a new training plan by the trainee is a learning outcome at this level.

4.2.1.6 Evaluation

Evaluation involves quantitative or qualitative judgments using internal evidence or external criteria. Unlike opinions which may not consider various aspects of the object, judgments are highly conscious and based on comprehension and analysis of the object (Bloom, 1956, 186–187). The criteria for judgments are well-defined. Internal evidence is concerned with consistency or accuracy. External criteria consider the utility of specific means for a given purpose and are often associated with standards of excellence in a field. A learning outcome at the evaluation level in the training of software engineers in human-computer interface, for example, may be to create and conduct a usability test for a software application.

4.2.2 THE AFFECTIVE DOMAIN—KRATHWOHL'S TAXONOMY

The affective domain is concerned with learning outcomes that describe changes in interests, attitudes, and values (Bloom, 1956, 7). Affective learning outcomes

Learning or Performance Objectives

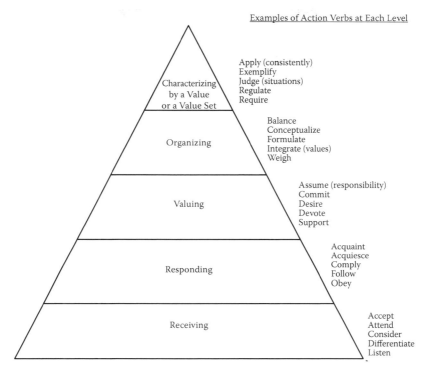

FIGURE 4.2 The affective domain and hierarchy of learning outcomes (Krathwohl, D. R. et al., 1956. *Taxonomy of Educational Objectives: The Classification of Educational Goals; Handbook 2 Affective Domain*. New York: David Mckay). Note that an action verb may be suitable for more than one level of learning outcome depending on the context.

consider feelings and emotions in addition to observable behavioral changes. For this reason they are difficult to describe or measure. Surveys or long-term observations of behavior change may be needed. Figure 4.2 presents a summary of the five levels of learning outcomes within the affective domain, with examples of action verbs that may be used. Similar to the cognitive domain, the levels reflect learning outcomes manifested in behaviors that range from the simple, concrete, and less pervasive to the complex, abstract, and pervasive (Krathwohl et al., 1956, 24). Some action verbs may be applicable to more than one level.

4.2.2.1 Receiving

Receiving is awareness of the existence of certain ideas, concepts, or phenomena. The individual may become simply aware of the information provided, or may learn to listen to controversial issues with an open mind to opposing viewpoints. As an example, in diversity training where the emphasis is the relationship of diversity to effectiveness and efficiencies in organizational performance, trainees may be introduced to the theories of dimensions of culture and their implications on management decisions.

4.2.2.2 Responding
Active participation begins at this level. The individual is somewhat committed to the idea and undertakes activities in the learning process to know more about the topic. Continuing with the example of diversity training, the trainees' degree of participation may range from reading assigned materials to researching other literature on issues of cultural diversity to gain a deeper understanding.

4.2.2.3 Valuing
The individual makes a decision on the value of the concepts and the extent of his or her involvement. This is the level where definite commitment is shown. Although values are internalized, clues to those values may be observed in overt behaviors. Learning outcomes are linked to behaviors that identify the values. The trainees in the diversity training may be willing to be perceived as supporting diversity in the workplace to promote utilization of the best talents.

4.2.2.4 Organizing
The individual integrates the value with his or her existing values, resolving any conflicts and building an internally consistent value system. The set of attitudes and beliefs in the value system are ordered taking into consideration the relationships among them. A learning outcome in the diversity training at the organizing level may be to adopt the theories and practices of managing cultural diversity and to formulate them into a work plan.

4.2.2.5 Characterizing by a Value or Value Set
The individual has internalized the value system and will apply it to other situations consistently. Over time the individual has manifested behavior compatible with the value or value set. The learning outcome emphasizes the fact that the intended behavior has become a characteristic of the individual. The trainees in the diversity training would be regarded by observers as definitely committed to maximizing organizational performance through managing cultural diversity.

4.2.3 THE PSYCHOMOTOR DOMAIN

The original committee of college and university examiners that constructed the taxonomies for the cognitive and affective domains did not create a taxonomy for the psychomotor domain. It did not believe that such a classification would be useful (Bloom, 1956, 7–8). Subsequently other authors published several taxonomies of the psychomotor domain, including Simpson's (1966) research report to the United States Department of Health, Education and Welfare, Dave's (1970) work *Psychomotor Levels* (as cited in Krathwohl, 2002), and Harrow's (1972) book *A Taxonomy of the Psychomotor Domain: A Guide for Developing Behavioral Objectives*. Practitioners at educational institutions have combined some of the classifications into modified versions of the psychomotor taxonomy. A version that is suitable for physical and kinesthetic skills commonly required in the work environment (Bixler, 2007),

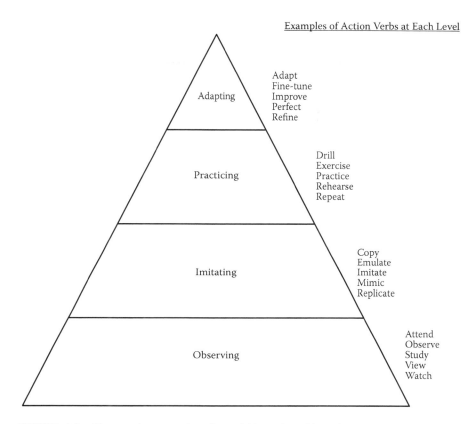

FIGURE 4.3 The psychomotor domain and hierarchy of learning outcomes (Bixler, B., 2007. "Psychomotor domain taxonomy," *Penn State Learning Design Community Hub.* http://ets.tlt.psu.edu/learningdesign/objectives/psychomotor). Note that an action verb may be suitable for more than one level of learning outcome depending on the context.

such as keyboarding and using technical instruments, is presented in Figure 4.3. The figure shows four levels of learning outcomes that progress from observation to mastery of the skill and examples of action verbs.

4.2.3.1 Observing

With or without supplemental training aids such as reading materials or audiovisual media, a trainer or other experienced individual performs a task while the trainee observes. The trainee is expected to pay special attention to the sequence in which the task is performed and the finished product; therefore, "active mental attending" of the event is involved (Bixler, 2007). For example, a trainee learning how to build an airplane model for display may be taught the materials and steps to be used and then observe someone demonstrate the process. While observing, the trainee would be relating what he or she sees to what has been taught.

4.2.3.2 Imitating

Provided with directions and sequences and under close supervision, the trainee imitates the behavior that he or she is to learn. The physical movements may not be smooth or coordinated since the trainee is just acquiring the rudiments of the skill (Bixler, 2007). In the above example, it would be the trainee's turn to build the model. The trainer will give directions and supervision to ensure, among other things, that the trainee does not seal up the fuselage of the model before placing the miniature seats inside!

4.2.3.3 Practicing

The trainee repeats the task again and again so that performance becomes habitual. The trainee acquires the skill and is able to perform with the correct sequence, timing, and coordination (Bixler, 2007). The trainee in the example practices building the model many times, to the point where the steps are performed naturally and correctly. Note that imitating and practicing may be done in the real or virtual environment depending on many factors, two of which are cost and time. For example, if the airplane model is of table-top size, the trainee may practice with actual materials every time as the cost is not prohibitive. If the model is a large display, part of the imitating and practicing is probably done virtually. Chapter 20 has more in-depth discussions on virtual training.

4.2.3.4 Adapting

This is the level where the trainee makes minor adjustments in the physical activity in order to perfect it, possibly with the help of the trainer or a mentor who would supply tips and tricks (Bixler, 2007). Practice makes perfect, and fine-tuning makes an individual an expert. For instance, in making the airplane model, special skills in applying the finishing touches could make the model stand out as an art form.

4.3 MEANINGFUL LEARNING OBJECTIVES

Before beginning to write the learning objectives, you would have performed the training needs analysis and the task analysis. The training needs analysis confirms that training is likely to result in higher performance and the task analysis reveals specifically what should be taught or emphasized in the training.

For instance, a task analysis has identified eight sequential steps in the preparation, performance, and follow-up phases of a task. Employees performing that task are doing every step correctly except steps 3 and 4. In such a case, the training may focus on those two steps and the learning objectives would be based on what the trainees should do at that point. It follows that there should be at least one learning objective for each task or topic that is covered in the training. The learning objectives should be described precisely since they are the cornerstone of the training plan.

Learning objectives should be listed from the simple to the complex or in the order in which the task is performed. They should state any prerequisite that must be met before or at the time of attending the training. The prerequisite could be something as simple as a requirement to bring a calculator to the training so that

Learning or Performance Objectives

the trainee can participate in exercises that involve computation. An example of a prerequisite for an advanced course for job incumbents might be "participants must have at least 12 months' on-the-job experience." These prerequisites affect how the audience and condition will be defined, as explained in the next section.

4.3.1 THE ABCD FORMULA

Every learning objective should state what the trainee should be able to do upon completion of the training. It should include four essential components (Heinich et al., 1999, 36–38):

Audience
Behavior
Condition
Degree

4.3.1.1 Audience

The target audience should have been characterized during the needs assessment and recorded in the first part of the training plan. At a minimum, you should know the trainees' job functions, experience, education background, gender, approximate age group, and if anyone in the group is not fluent in the language to be used for the training. The learning objectives and contents of training on the same topic would be quite different for an audience that is new to the job, compared with an audience that has years of experience. If there is a language barrier, you may need to apply special techniques as discussed in Chapter 16. With regard to accommodation, usually it is sufficient to include in the training announcement the contact information for anyone in need to submit a request in advance of the scheduled training.

4.3.1.2 Behavior

What do you want the audience to do or what is the trainee behavior that you will attempt to change with the training? The answer to this question guides you in the choice of action verbs in the learning objectives. The examples of action verbs given in Figures 4.1, 4.2, and 4.3 are by no means exclusive, nor are the levels of learning outcomes at which they are placed carved in stone. Some action verbs can be used for more than one domain of learning objectives or more than one level within a domain. In particular, in the psychomotor domain many words directly describing the action may be used, such as "assemble," "load," or "operate."

4.3.1.3 Condition

What is the condition under which the audience will demonstrate its ability to perform the desired behavior? The condition may be a constraint such as time limit, or if resources will be provided. It may influence whether a trainee meets the criteria for successful completion of the training course. Suppose after the training, trainees are expected to perform a task with all the steps in the correct sequence. A trainee who is able to do this using a checklist may not be able to do it without the checklist. The condition, whether a checklist can be used, must be stated in the learning objective. Another example is when a trainee must complete a task within a certain amount of time.

4.3.1.4 Degree

What level of performance will be acceptable? The degree of mastery may be expressed in terms of quality, quantity, time, cost, accuracy, consistency, and other parameters. The degree applied in the training should be the same as or closely related to the expected job performance. For agents being trained in outbound sales, the degree may be the number or dollar amount of sales closed. In training employees in the fast-food industry, the degree may be the amount of time to prepare a particular menu item. In a written test, the degree may be the percentage of correct answers.

4.3.1.5 Examples

The following examples illustrate how learning objectives may be structured with the "ABCD" formula. Different font styles are used to indicate the parts of each objective that are related to the audience (underlined), behavior (italic), condition (bold), and degree (bold italic).

- Cognitive—knowledge: The medical assistants *will be able to recite the steps of patient assessment* **from memory** ***with 100% accuracy.***
- Cognitive—application: **Based on information provided by the client,** the tax preparers *will be able to identify and explain to the client the applicable forms and schedules for a federal income tax return* ***within three minutes.***
- Affective—responding: The managers *will explore the power of emotional intelligence* **by reading additional materials on their own** ***within a week after the initial training.***
- Affective—characterized by value: The rehabilitated persons *will refrain from illegal drug use* **all the time** **whether they are alone or in the company of others.**
- Psychomotor—imitating: The trainees in emergency response *will put on the self-contained breathing apparatus* **using the exact method as the trainer** **while they follow the trainer step-by-step.**
- Psychomotor—practicing: The new operating engineer *will be able to perform emergency generator testing* **according to standard operating procedures** **without referring to the user manual.**

4.3.2 THE SMART PRINCIPLE

The SMART acronym is used frequently in goal setting, emphasizing the importance of having goals that are specific, measurable, action-oriented, realistic, and time-bound. Similarly, learning objectives should be SMART, with a slightly different meaning:

Specific
Measurable
Actionable

Learning or Performance Objectives

Results-oriented
Trainee-centered

4.3.2.1 Specific

A learning objective should state clearly what the trainee will be able to do after the training; otherwise misgivings may arise as to expectations. For example, a poorly written learning objective may use a vague term such as "familiarize." More specific verbs could be "describe" or "identify." Lack of specificity means that the outcome will be difficult to measure.

4.3.2.2 Measurable

A learning objective should reference an outcome that is measurable and observable. When an outcome can be measured and observed, the performance standard and the testing are less likely to be subjective and biased. Earlier discussions have pointed out that learning outcomes in the affective domain are hard to observe as they encompass changing values. Granted that some learning outcomes are easier to observe than others, all of them should be measurable directly or through an appropriate surrogate (Johnston, 1999).

4.3.2.3 Actionable

A learning objective should have an action verb that relates to the behavior. A learning objective can be intermediate or final; thus, the action verb sometimes reflects part of the learning process instead of the final outcome.

Consider a leadership training workshop on negotiation skills that has an intermediate and a final learning outcome:

- Intermediate: Without referring to the course materials, participants will discuss all the reasons why a win-win situation is desirable.
- Final: In a 15-minute role-playing, each pair of participants will negotiate a hypothetical business contract to arrive at a win-win situation, based on scenarios to be provided by the trainer.

Whereas the intermediate learning objective serves as a check for comprehension of the advantages of a win-win situation, the final learning objective reveals that the trainer wants the trainees to demonstrate their ability to use the negotiation skills learned. Comprehension is part of the learning process before the application level is accomplished.

4.3.2.4 Results-Oriented

The ultimate purpose of training is to transfer knowledge, teach skills, or transform behavior in support of the achievement of an organization's goals. When writing a learning objective, always keep in mind the results to be accomplished—the trainees' demonstration of having acquired the knowledge, skills, or behavior after the training course.

4.3.2.5 Trainee-Centered

Learning objectives are trainee-centered because they are statements of what the trainees will be able to do as a result of what they learn in the training. The objectives should not describe what the trainer plans to do during the training event. This aspect of learning objectives is probably the one that incidental trainers most often forget. You can overcome this challenge with practice and by applying the "ABCD" formula and the "SMART" principle.

4.4 NEXT STEPS

Having laid the cornerstone of your training plan, you are ready to draw up the rest of the plan, beginning with the instructional strategies!

REFERENCES

Bixler, B. 2007. "Psychomotor domain taxonomy," *Penn State Learning Design Community Hub.* http://ets.tlt.psu.edu/learningdesign/objectives/psychomotor.

Bloom, B. S., ed. 1956. *Taxonomy of Educational Objectives: The Classification of Educational Goals; Book 1 Cognitive Domain.* New York: Longman.

Dettmer, P. 2006. "New blooms in established fields: Four domains of learning and doing." *Roeper Review* 28 (2): 70–78.

Harrow, A. J. 1972. *A Taxonomy of the Psychomotor Domain: A Guide for Developing Behavioral Objectives.* New York: David McKay.

Heinich, R., M. Molenda, J. D. Russell, and S. E. Smaldino. 1999. *Instructional Media and Technologies for Learning.* Upper Saddle River: Prentice-Hall.

Johnston, K. C. 1999. "What are surrogate outcome measures and why do they fail in clinical research?" *Neuroepidemiology* 18 (4): 167–173.

Krathwohl, D. R. 2002. "A revision of Bloom's taxonomy: An overview." *Theory Into Practice* 41 (4): 212.

Krathwohl, D. R., B. S. Bloom, and B. M. Masia. 1956. *Taxonomy of Educational Objectives: The Classification of Educational Goals; Handbook 2 Affective Domain.* New York: David Mckay.

Simpson, E. J. 1966. *The Classification of Educational Objectives, Psychomotor Domain* [Report Number BR-5-0090]. Washington, DC: United States Department of Health, Education and Welfare.

5 Instructional Strategies

5.1 INSTRUCTIONAL STRATEGIES DEFINED

Instructional strategies are an essential part of any training project. Some authors use the term to include everything in the training plan, from developing the course materials to choosing the assessment method. Others refer to instructional strategies as the modalities in which content is delivered and trainees are engaged in activities. It is in the latter sense that the term is used in this book.

5.2 THE UBIQUITOUS LECTURE

Visualize this scenario: A project manager is conducting a training course in an auditorium. Standing behind a lectern, he delivers a two-hour, nonstop lecture on project management. Thirty minutes into the "training," one-third of the trainees are struggling to stay awake, with another one-third actually dozing off. The others are surfing the Internet on their computers and tablets or texting on their smartphones. For those conscientious trainees who try to listen, at best they would remember 20% of what they heard.

The above description is not an exaggeration. In fact I was at that event and the project manager is a friend of mine. Unfortunately this kind of situation happens more frequently than one would think because many incidental trainers are unaware of, or uncomfortable with, using instructional strategies besides lecture. Furthermore, lectures are overused since it is the easiest strategy to implement. That does not mean it is the most effective.

So, what is the most effective? There is not one answer.

5.3 THE "CONE OF EXPERIENCE"

In his classic work *Audio-Visual Methods in Teaching*, Dale (1946, 37–52) describes the "cone of experience" that exemplifies the connection between different instructional methods in enhancing the learning experience. They range from the more concrete to the more abstract ways of delivering course materials, that is, from purposeful experiences to demonstrations, field trips, exhibits, audiovisual materials, and visual and verbal symbols. According to Dale, direct experience is the basis of all effective learning (38). Since the degrees of sensory experience differ among these instructional strategies, the choice of a strategy depends on how much sensory experience the trainer desires to provide the trainees, which in turn is dependent on the subject matter and the audience. In addition, a multimodal approach helps to enrich the learning experience and aid trainees in developing concepts from experiences.

Since its publication, the "cone of experience" has been modified by other individuals and entities. They have related instructional strategies to varying degrees of learning retention, such as "we retain 10% of what we read, 20% of what we hear," and so on. As of this writing, a search of the literature and inquiries with esteemed colleagues in the field of education have not revealed empirical research that directly supports these percentages. However, they demonstrate a consensus that audiovisual experience greatly enhances learning retention.

5.4 LEARNING STYLES

Instructional strategies have also been associated with learning styles. A person's learning style is the cognitive, affective, and psychological make-up that influences how the individual interacts with environmental, emotional, sociological, physical, and psychological stimuli in the learning process (Dunn, 1984). Learning style theories claim that such personal characteristics affect which instructional strategies are most conducive to a particular individual's learning, hence learning styles are also referred to as learning preferences.

There are many models of learning styles and instruments that help to determine an individual's learning style. According to Curry's 1987 book *Integrating Concepts of Cognitive or Learning Style: A Review with Attention to Psychometric Standards* (as cited in Hickcox, 1995, 28–29), these instruments may be organized into three levels:

- Instructional and environmental preferences
- Information processing preferences
- Personality-related preferences

Incidental trainers would be most concerned with the first level, the instructional and environmental preferences, which is also the most observable level. One popular model of learning preferences, developed after Curry's review, is the VARK model (Fleming and Mills, [1992] 2011). It describes learners' perceptual modes as visual (V), aural (A), read/write (R), and kinesthetic (K). The visual learner has a "preference for graphical and symbolic ways of representing information," such as pictures and charts. The aural learner responds better to auditory stimuli, as when listening to lectures and discussions. The learner whose preference is to read/write favors materials printed in words, as in a dictionary or glossary. The kinesthetic mode is "the perceptual preference related to the use of experience and practice (simulated or real)." Although "experience and practice" may involve the use of other perceptual modes such as sight or touch, the emphasis is on connecting the learner to "reality" through experiment or simulation.

Learning style theories are based on the hypothesis that individuals learn and remember best if the material is delivered to them in formats that match their learning styles (e.g., Dunn, 1984). In other words, to be most effective an instructional strategy should be aligned with the learning preference, that is, visual presentation should be given to the "V" learners, oral instruction should be delivered to the "A" learners, and so on. This is called the meshing hypothesis. However, studies and reviews have challenged this hypothesis. For example, Krätzig and Arbuthnott (2006) find that their study participants' self-report of learning styles may not be

Instructional Strategies

accurate because the participants answered the learning style inventory questions using general memories and beliefs that were not necessarily related to learning in specific modalities. Coffield et al. (2004, 139) identified 71 learning style models and examined the most influential 13 for internal consistency, test-retest reliability, and construct and predictive validity. They conclude that only three of the 13 models could meet such criteria. Pashler et al. (2008) find that the validity of learning styles as applied to education has not been sufficiently tested using appropriate experimental methodology. Both the Coffield et al. (2004, 2, 119, 128, 138, 145) and Pashler et al. (2008) reviews also point out the commercialization of the learning style instruments whereby the publishing and selling of these instruments have grown into an enormous amount of business activities and profits for the creators, resulting in a lack of incentive for independent research to validate the instruments.

Where does that leave the incidental trainer?

5.5 MANY ROADS, ONE DESTINATION

First, the important thing to remember is that your goal as a trainer is to help your trainees achieve their learning objectives. All instructional strategies you use should support this goal. Second, as Kneale et al. (2006, 126) indicate, individuals in a class will have different learning styles. Unless the training is one-on-one, as in some on-the-job or virtual training situations, a group of trainees probably will consist of persons with different learning styles, no matter which learning style model is used to define the learning styles. This means that based on the learning style theories, one instructional strategy will be insufficient to accommodate all the trainees simultaneously and a better approach is to employ multiple strategies (Ross et al., 2001, 410). Third, Willingham (2005) states that it is not the learner's best modality but the content's best modality that counts. Some contents are better taught with visual images while others are better taught with verbal expressions, and many topics need to be presented in more than one modality. Fourth, according to the "cone of experience" (Dale, 1946, 42), a multimodal approach is beneficial because it helps each trainee develop concepts from experiences. Fifth, the attention span for adults is estimated to be no more than 20 minutes (Murphy, 2008). Moving from one instructional strategy to another helps you maintain audience interest and attention. For these reasons, the best advice in training design is to use a variety of instructional strategies.

Instructional strategies can be grouped into the following general categories. They can be used individually or in combination. Their effectiveness is situational, depending on the audience and subject matter of the training.

- On-the-job training
- Lecture and panel
- Group discussion
- Demonstration and practice
- Role-playing
- Self-guided discovery
- Collaborative learning

5.5.1 ON-THE-JOB TRAINING

It may not occur to some incidental trainers performing on-the-job training that they are "trainers," but that is their role. On-the-job training may take the form of providing instructions on how to perform a task or job, or the trainer-trainee may have a closer interpersonal relationship as mentor-mentee or "buddies," whether formal or informal. Such training circumstances may arise between a supervisor and his or her direct report, or a senior colleague and a new employee. Mentoring requires counseling skills in addition to training skills and is beyond the scope of this book. Job instruction training is often concerned with the performance of a physical task and consists of four steps:

1. Preparation: The trainer determines what the trainee should know and finds out what the trainee already knows. This amounts to an informal needs assessment.
2. Presentation: The trainer explains and demonstrates performance of the task, one step at a time.
3. Performance: Being observed by the trainer, the trainee performs the task while explaining the key points to show understanding. This process is repeated until the trainer is satisfied that the trainee knows how to do the task.
4. Follow-up: The trainee performs the task on his or her own and the performance is monitored. The trainer or a designated person is available for the trainee to ask questions until the trainee no longer needs help.

On-the-job training is most suitable for achieving application-level learning objectives or developing problem-solving skills as it has the advantage of demonstration and practice in real conditions. The learning outcome is immediately apparent. Any deficiency in the trainee's performance can be corrected on the spot. A trainee receives personal attention, which may be a cause for higher motivation. The disadvantages of on-the-job training are that the contents provided by different trainers for the same task are seldom standardized and, more important, the trainer may be selected only because the person is available, not because he or she has training skills or the best experience in the particular task or job.

5.5.2 LECTURE AND PANEL

Although the lecture format usually includes a question-and-answer period after a formal presentation, it is primarily one-way communication from the trainer to the trainees. The trainer presents concepts or facts with or without the use of training aids such as slides.

Occasionally a trainer may invite a guest speaker to deliver a lecture, to benefit from the in-depth knowledge or experience of an expert on a particular topic. Guest speakers can offer different perspectives and reinforce learning. Frequently, they are willing to give advice or answer questions from trainees after the training event. Having guest speakers is especially beneficial during long training events that last for more than a few hours as it adds variety. Expertise of the guest speaker is a major consideration, but if you invite a guest speaker, you should also make sure that the guest speaker is a good trainer and understands the target audience. You must coordinate with the guest speaker what areas to cover to avoid duplicating other materials in the training course. It is desirable to request the guest speaker to provide the contents of his or her portion of the training for your review prior to the training event. Agree upon and adhere to time limits.

One way to add even more variety and viewpoints is to have a panel. The size of the panel depends on the time available and usually consists of no more than five panelists. Each panelist gives a lecture and five to ten minutes are allowed for questions and answers after all the presentations. As in the case of a single guest speaker, you should be clear as to what each panelist will present and ensure beforehand that the contents are relevant and do not overlap. During the panel discussion, the audience may direct a question to a specific panelist, or more than one panelist's opinions may be sought. Your role is that of a moderator. It is a good idea, though, to prepare a few "seed questions" to get the ball rolling in case no one from the audience submits a question, and to draw out the expertise of the panel that is most relevant to the learning objectives.

A lecture is appropriate when the learning objective is at the knowledge level. It offers a convenient way to introduce a topic at the beginning of a training course to familiarize trainees with theories, concepts, or facts. For example, a trainer may kick off a training course on how to conduct air quality assessment with a lecture describing the laws and regulations on air quality.

A main advantage of lecture is that the primary trainer or guest speakers can retain control with ease and can present a lot of information in a relatively short time to a large group of trainees. A lecture can be entertaining—which helps information retention—if the presenter is enthusiastic and engaging. Unfortunately, oftentimes lectures are boring and do not stimulate trainees to develop reasoning skills.

5.5.3 Group Discussion

Group discussions are particularly suitable for comprehension-level or problem-solving learning objectives. They should not be casual conversations. Rather, they should be structured toward the learning objectives. Using the previous example of a training course on how to conduct air quality assessment, a proper discussion might center on how the laws and regulations apply to different types of situations. On the other hand, a political debate on whether such laws and regulations should have been passed is irrelevant. It is up to the trainer, as a facilitator, to bring the group back when it deviates from the correct focus.

Group discussions may be organized for the whole group of trainees or subgroups during a breakout session. The latter is also called a buzz session. Whether used with the whole group or subgroup, a round robin format can ensure that each member of the group, going around the table in a clockwise or counterclockwise direction, has an opportunity to express ideas. Group discussion can also take the form of brainstorming, which encourages everyone to think spontaneously and contribute out-of-the box solutions to a problem, on the basis that all ideas will be considered and no idea is stupid.

By encouraging input from trainees, especially those who already have knowledge or experience in the topic, a guided group discussion stimulates interest and participation. This modality allows you to check for understanding before the training ends by asking pertinent questions and listening to what the trainees have to say. To facilitate a productive group discussion, stay focused on the learning objectives, encourage everyone to participate, and do not allow a few people to dominate the discussion.

5.5.4 Demonstration and Practice

Demonstration and practice can occur on the job or in the classroom. This strategy is best used for application-level or psychomotor learning outcomes and is suitable for many types of tasks. For instance, a trainer may show trainees how to operate or troubleshoot a piece of equipment, followed by the trainees' performance of the same task. In software application training, a trainer may set up a demonstration of how to use certain features of the program and let the trainees practice the procedures on their computers. Since the demonstration is a model for the trainees to follow, the trainer should have determined all the correct steps in the right sequence prior to the training. The person doing the demonstration should follow the exact procedure and be proficient in performing the task. This person can be the trainer or someone experienced in the job. If an experienced employee is recruited to help with the training, ensure that the individual will demonstrate according to the standard, since over time some experienced employees may develop their own ways of doing things. If the modification is acceptable or desirable, it should be incorporated into the standard procedure.

An obvious advantage of demonstration and practice is that the trainees can gain hands-on experience. Research has shown that hands-on experience is superior to demonstration alone in achieving problem-solving learning objectives (Glasson, 1989; Korwin and Jones, 1990). Such experience is essential for psychomotor learning objectives like troubleshooting equipment malfunction. This strategy also has the advantage of engaging multiple senses. The main disadvantage is that the preparation and delivery can be time-consuming. Equipment or model needs to be set up and, depending on its size and complexity, the logistics can be involved. Safety risks may arise, although these can be minimized by the application of virtual reality and simulation in training.

5.5.5 Role-Playing

Role-playing is another method of learning by doing. It is similar to demonstration and practice as it gives a trainee an opportunity to experience a situation first-hand. It is different from demonstration and practice in that it can be used for learning objectives in the affective domain. For this purpose, the trainees can be asked to put themselves in another's shoes, to see things from a different or opposing perspective. The strategy works well too when trainees must learn how to resolve complex issues or to apply skills learned, such as coaching or negotiation skills.

Instructional Strategies 41

In role-playing, the trainer sets up a scenario or simulated event, which should be based on an actual situation. One or more trainees act out the characters in the event. For example, in a training course on negotiation skills, the event may be a real estate transaction. One trainee plays the part of the buyer and another plays the part of the seller. The two would use the negotiation skills taught in the course to negotiate the price of a piece of property. If time permits, they can reverse the roles after the first round. A group discussion can ensue to analyze how effective the players have used the negotiation techniques and if another approach may be an improvement. If this type of group evaluation is done, the trainer should set ground rules for comments to be constructive feedback and not criticism.

Role-playing embraces creativity and is conducive to building problem-solving skills. Its utility for affective learning objectives lies in the ability to help trainees see another viewpoint, which may alter their original perception or opinion. It is harder for the trainer to control time in role-playing than, say, in a lecture. If you employ this modality, set time limits and strive to stay on schedule. As discussed in Part 2, some trainees may feel uncomfortable in role-playing situations due to cultural influences or personal characteristics. Be sensitive to such issues.

5.5.6 SELF-GUIDED DISCOVERY

Self-guided discovery can be implemented in numerous ways. In general, it follows the pattern of inquiry training (Reed, 2003, 119–121). The trainer presents a problem to the trainee, who gathers data from the course materials or elsewhere and formulates a solution to the problem. A final phase requires the trainee to reflect on the process and find ways for future improvement.

This strategy is often used with technology-oriented training platforms such as computer based training (CBT), e-learning, and mobile learning. Asynchronous course delivery using these platforms relies heavily on self-guided discovery since the course is akin to a self-study program. However, as long as trainees are to arrive at conclusions based on information given or new information obtained from their own research, they are developing analytical skills through the process of self-guided discovery, whether they are in the virtual training environment.

With proper design and preparation, self-guided discovery is especially effective for building problem-solving and leadership skills. It has the advantages of being trainee-centered and advancing trainees' reasoning ability. It can be self-paced. One condition of self-guided discovery is that trainees must be able to work independently. This aspect can be an advantage when the trainees are motivated. It becomes a disadvantage when they are not. In addition, the strategy may be unsuitable for novices as basic knowledge and skills on the topic or job are needed for a trainee to enjoy and learn through the discovery experience.

5.5.7 COLLABORATIVE LEARNING

Self-guided discovery in a group setting is collaborative learning. Trainees work together in small groups, usually between two and five members, to accomplish a task or achieve a shared goal. The size of the group would depend on the aim.

A larger group is desirable for generating ideas while a smaller group is better for working on differentiated tasks (Reeve and Shumway, 2003, 135).

Breakout sessions in training are excellent opportunities for collaborative learning. Each group is assigned a subtopic or problem to work on. Members of the group work together and when finished, one person reports the conclusions or solutions on behalf of the group. A variation of this approach is to have each member within a group work on a specific part of the problem. For example, in trying to find solutions to sustainability, one member might research ways to recycle paper and another member might study methods of recycling electronic waste. Each member will share the information with the group. Collaborative learning can also take the form of debate and training games. In a debate, trainees are divided into teams that are assigned to take positions in favor of or against a controversial issue. Team members research the issue and present arguments supporting their side. Games are described in more detail in the next chapter on training aids and media.

As in individual self-guided discovery, collaborative learning promotes the development of analytical and reasoning skills. A disadvantage, however, is that to be successful it also requires teamwork and social skills, which some trainees may not have. Situations may arise where some members of a group do not contribute substantially to the group effort. Careful planning is advisable before implementation of a collaborative learning activity.

5.6 SELECTION OF THE "BEST" STRATEGIES

Which strategies are the best? It depends on a number of factors. The issues to consider are learning objectives, target audience, trainer's skills, and situational constraints.

5.6.1 LEARNING OBJECTIVES

As the foundation of the whole training plan, learning objectives should be the first and foremost consideration in determining what instructional strategies to apply. Match the instructional strategies to the type and level of learning objective. As learning objectives progress from knowledge level to application and problem-solving level, you can change instructional strategies accordingly. For example, the project management training described in Section 5.2 could have employed a combination of lecture, demonstration and practice, and group discussion. The project manager could begin the training with a brief lecture on the purpose and process of project management, demonstrate the techniques of using project management tools, let the trainees practice using the tools, and lead a group discussion on how each trainee might apply the techniques in their real-life work projects.

5.6.2 TARGET AUDIENCE

Learning style theories argue that trainers should consider the audience's learning styles when selecting instructional strategies. Besides the VARK types, other labels such as "reflective observers" and "active participants" are given to learners. It is

Instructional Strategies 43

said that a strategy suitable for one type of learners may be unsuitable for another type. As mentioned earlier, when the audience is a diverse group, implementing more than one strategy is crucial to maximizing learning for more reasons than one. Furthermore, the most effective trainer is prepared to be flexible in switching strategies on the spot as he or she observes audience reactions.

There are other audience characteristics that you should think about when planning instructional strategies. Of particular relevance are current knowledge, skills, and abilities (KSAs), as well as generational and cultural differences within the group. The KSAs would have been linked to the learning objectives. Trainees in a basic course may require proportionately more lecture to be introduced to the topic. Trainees in an advanced course would find the same amount of lecture uninteresting; instructional strategies such as team activities can be used to elicit experiences from the advanced trainees so that best practices can be shared. Generational and cultural differences are discussed in Chapters 15 and 16, respectively.

5.6.3 Trainer's Skills

Some instructional strategies demand a higher skill level on the part of the trainer than others. One reason why lectures are overused is because most trainers are comfortable with it and feel they are in control. It is easy for unskilled trainers to lose control over group discussions. Practice and role-playing require good time management. With self-guided discovery and collaborative learning, the trainer must articulate clearly what information the trainees must research or what problem they are trying to solve.

You want to employ strategies you feel comfortable with in order to use them smoothly and effectively. At the same time, improving your proficiency in handling other strategies will stretch your comfort zone and expand your choices. Colleagues and peer groups can be extremely helpful in providing suggestions for improvement as you prepare and practice new strategies.

5.6.4 Situational Constraints

Factors beyond your control may limit your choices of instructional strategies. Constraints may relate to group size, available time, logistics, room setup, and venue. For instance, it may be impossible to offer every trainee hands-on practice if only one set of needed equipment is available for 30 trainees during a one-hour training session. An auditorium may be the only venue available but it is not ideal for small group discussions. Whereas on-site demonstration provides a valuable practical experience, the location may not have the facility for a lecture if that is also desired. Potential constraints underscore the importance of advance planning in selecting training aids, media, and venue. These are the topics of the next two chapters.

5.6.5 Summary

Table 5.1 presents a quick reference of how instructional strategies are matched with objectives at different levels of learning outcomes.

TABLE 5.1
Matching Instructional Strategies with Learning Objective Levels

Instructional Strategy	Learning Objective Level
On-the-job training	Application, problem-solving
Lecture and panel	Knowledge
Group discussion	Comprehension, problem-solving, evaluation
Demonstration and practice	Application, psychomotor, problem-solving
Role-playing	Affective, application, problem-solving
Self-guided discovery	Problem-solving, analysis, synthesis, evaluation
Collaborative learning	Problem-solving, analysis, synthesis, evaluation

REFERENCES

Coffield, F., D. Moseley, E. Hall, and K. Ecclestone. 2004. *Learning Styles and Pedagogy in Post-16 Learning: A Systematic and Critical Review*. London: Learning and Skills Research Centre.

Dale, E. 1946. *Audio-Visual Methods in Teaching*. New York: Dryden Press.

Dunn, R. 1984. "Learning style: State of the science." *Theory Into Practice* 23 (1): 10–19.

Fleming, N. D., and C. Mills. (1992) 2011. "Not another inventory, rather a catalyst for reflection." First published in *To Improve the Academy* 11: 137–155. Accessed August 24. http://www.vark-learn.com/documents/not_another_inventory.pdf.

Glasson, G. E. 1989. "The effects of hands-on and teacher demonstration laboratory methods on science achievement in relation to reasoning ability and prior knowledge." *Journal of Research in Science Teaching* 26 (2): 121–131. doi:10.1002/tea.3660260204.

Hickcox, L. K. 1995. "Learning styles: A survey of adult learning style inventory models." In *The Importance of Learning Styles: Understanding the Implications for Learning, Course Design, and Education*, edited by R. R. Sims and S. J. Sims. Westport: Greenwood Press, pp. 25–47.

Kneale, P., J. Bradbeer, and M. Healey. 2006. "Learning styles, disciplines and enhancing learning in higher education." In *Learning Styles and Learning: A Key to Meeting the Accountability Demands in Education*, edited by R. R. Sims and S. J. Sims. New York: Nova Science Publishers, pp. 115–128.

Korwin, A. R., and R. E. Jones. 1990. "Do hands-on, technology-based activities enhance learning by reinforcing cognitive knowledge and retention?" *Journal of Technology Education* 1 (2): 26–33.

Krätzig, G. P., and K. D. Arbuthnott. 2006. "Perceptual learning style and learning proficiency: A test of the hypothesis." *Journal of Educational Psychology* 98 (1): 238–246.

Murphy, M. 2008. "Matching workplace training to adult attention span to improve learner reaction, learning score, and retention." *Journal of Instruction Delivery Systems* 22 (2): 6–13.

Pashler, H., M. McDaniel, D. Rohrer, and R. Bjork. 2008. "Learning styles: Concepts and evidence." *Psychological Science in the Public Interest* 9 (3): 105–119. doi:10.1111/j.1539-6053.2009.01038.x.

Reed, P. A. 2003. "Inquiry in technology education." In *Selecting Instructional Strategies for Technology Education*, edited by K. R. Helgeson and A. E. Schwaller. New York: Glencoe/McGraw-Hill, pp. 117–129.

Reeve, E. M., and S. Shumway. 2003. "Cooperative learning in technology education." In *Selecting Instructional Strategies for Technology Education*, edited by K. R. Helgeson and A. E. Schwaller. New York: Glencoe/McGraw-Hill, pp. 131–145.

Ross, J. L., M. T. B. Drysdale, and R. A. Schulz. 2001. "Cognitive learning styles and academic performance in two postsecondary computer application courses." *Journal of Research on Computing in Education* 33 (4): 400.

Willingham, D. T. 2005. "Do visual, auditory, and kinesthetic learners need visual, auditory, and kinesthetic instruction?" *American Educator* 29 (2): 31–35, 44.

6 Training Aids and Media

6.1 THE DOUBLE-EDGED SWORD

Training aids and media can be entertaining, reinforce learning, and help to engage the audience during a training event. They support various learning preferences. Brain research suggests that some people learn better with the left brain (words, logic, analytical thinking) and others learn better with the right brain (pictures, music, spatial relationships). Information that is presented with multiple media activates more senses and increases learning retention. For example, incorporating graphics along with text in a slide presentation will be more effective than using graphics or text alone.

The use of aids and media, however, is a double-edged sword. It presents a challenge to many incidental trainers. The most effective materials such as costume and games take much time to plan and prepare. Since incidental trainers have other responsibilities besides training, most are not skilled in and have no time to practice using these materials, resulting in undesirable effects and embarrassing situations.

6.2 UNLIMITED CHOICES!

Training aids and media can take many forms, from a simple sheet of handout to complex models and computer software. There are also many variations within each type, only limited by your imagination and creativity. The key to effective training using aids and media is that they must support the training goal.

The following sections discuss several common categories:

- Handout
- Slide presentation
- Video/audio
- Easel pad, dry erase board, electronic copyboard
- Model, prop
- Costume
- Game
- Computer, Internet, simulator

6.2.1 Handout

Most trainees seem to prefer to have handouts that they can refer to during or after the training. To save printing and handling costs, handouts are distributed electronically before or after the training. Occasionally, hard copies are provided at the training event for the convenience of attendees or when the trainer desires to lessen the possibility of dissemination of an electronic handout to people who are not attendees. Hard copies of supplemental material may also be given out.

The most common type of handout is a copy of the slide presentation used by the trainer during the training. Just as the slide presentation is overused, this handout format is overused because it is easy to produce, and inexpensive if distributed electronically. Many trainees have become so accustomed to having a copy of the slides that they automatically ask for it, whether they will ever refer back to it afterward. On the flip side, some feel that this type of handout is not helpful as it does not summarize the important points in complete sentences (Brier and Lebbin, 2009, 355).

If the handout is distributed in advance, it raises the question: Will it discourage trainees from attending the training or encourage them to pay less attention if they attend, since they can read the handout? One way to overcome this is to modify the handout version of the slides. For example, the actual presentation in a crime investigation training course may show a photo or video of a crime scene, used for group discussion, followed by another slide listing the suspicious items trainees should have noted from the scene and pointed out in their discussion. The slide with the answers can be removed from the handout version. Another way is to blank out keywords in the handout version, so trainees must think and take notes to fill in the blanks during the training. Slide presentation software such as Microsoft® PowerPoint® features an area where the author can compose notes for his or her own reference. Most presenters use this feature to some extent. The notes should be removed from the handout version.

On the other side of the coin, a handout may be useful as pretraining reading to prepare trainees for what they will be learning. If they do their "homework," the training can be carried out more efficiently. Handouts consisting of complementary materials such as manuals in electronic or paper format provide added value as a source of reference. Workbooks can be used for exercises during or after the training. Supplemental material can be included on special topics that may not be of general interest among all trainees.

The disadvantage of using handouts is that some trainees may not read the materials, while others may be more focused on reading the handout during the training than on listening, observing, and participating. If a lot of materials is disseminated as hard copies, the costs of printing and time can be high. Preparing a special handout or a modified version of a slide presentation takes extra time.

6.2.2 Slide Presentation

Slides are usually created with Microsoft PowerPoint or Corel® Presentations™. These programs offer multimedia features that integrate graphics, photographs, music, and video that can enhance viewer attentiveness and enrich the learning experience. As the old saying goes, "A picture is worth a thousand words." A trainer can talk about how devastating a tornado is, but the talk will not be as impactful as photographs or videos of bent steel towers and decimated homes after a tornado. Needless to say, the media should be relevant to the message, which should be relevant to the learning objectives. The media is a tool and should not overshadow the message.

Training Aids and Media

One caveat relates to the design of the slides. It is amazing how often trainers and presenters are unaware of or simply ignore the best practices in slide design. They seem to forget that creating slides is not just transferring text or a spreadsheet to a different file format. Based on human factors and ergonomics principles and research, Durso et al. (2011, 5–6) have these recommendations:

- Fonts: Use a sans serif font such as Tahoma, Arial, or Verdana, with a minimum font size of 22-point for bullets and 16-point for figure legends and axes.
- Colors: Use high contrast between the text and background colors, for example, dark text with light background and vice versa. The dark text–light background combination is preferred under all lighting conditions. Avoid red–green combinations as 5%–8% of the male population has color deficiency, most commonly red–green color blindness.
- Layout: Keep font and color consistent throughout the presentation. Line spacing should be half the height of a character. Leave a margin on all sides of the slides so that important information will not fall off the projection screen.
- Comprehension: Present a single main idea per slide, using three to five bullet points and the active voice. Do not write paragraphs of text.
- Charts, graphs, and tables: Avoid three-dimensional graphs or too many colors and shapes. Using texture and pattern in addition to colors helps those who have color deficiencies. It is preferable to place the legend close to the data than on the side.

Szul and Woodland (1998) suggest using a headline and a large first letter to capture attention, using uppercase and lowercase letters, and framing text or graphic for emphasis. Do not use italics.

The following are additional helpful tips:

- Use high-quality graphics: Many images and videos copied or downloaded from the Internet, even if legitimately used, have resolutions that are too low for projection on a large screen. The color and quality of most photographs taken with smartphones or tablets are inferior to those taken with digital cameras, which have a wide range of pixel count, image sensor, and other specifications. The resolution and quality of projection equipment also vary but you are always better off starting out with a high-quality image.
- Use standard RGB (red–green–blue) colors: Colors are rendered differently on different monitors and projectors. If customized colors are used, there are more chances that the colors on the screen will not appear as intended. If any artwork is originally created for the print medium, it is probably in the CMYK (cyan–magenta–yellow–black) color mode. Be sure to convert it to the RGB mode to ensure that the colors will appear vibrant on the screen. RGB is what projectors use.
- Use standard fonts: Standard fonts such as Arial are preferable to special fonts when you deliver the presentation with a computer different from the

one you have used to prepare the slides. The presentation computer may not have the special fonts, in which case it will replace an unavailable font with a standard font and possibly interfere with the layout if the size of the font is changed. The same happens when electronic handout is distributed, unless the distribution file is packaged with the correct fonts.
- Leave plenty of empty space: What is the point of putting up a "busy" slide and then saying to the trainees, "You probably cannot read this"? If the trainees are concentrating on trying to read the slide, they are not listening to you. Some trainers use the "7 × 7" rule, meaning no more than seven words per line and no more than seven lines per slide. Others reduce the amount of text to "6 × 6" or "5 × 5" or eliminate text altogether and use graphics instead.
- Vary the layout of the slides: The purpose is to add interest. For instance, use a text box on the left and graphics on the right on one slide and reverse the sides on another. However, avoid "jumping around" too much as that can be distracting.

Another important point to remember is copyright and privacy. Copyright ownership pertains to the use of all course materials, but it should be specially noted with regard to the use of graphics and videos. There is sometimes a misconception that any photo, clip art, or video downloaded from the Internet can be used freely. When incorporating media created by others, it is necessary to make sure that any required permission is obtained. Permission may also be needed when using photos with recognizable faces or property, depending on the circumstances under which the pictures were taken and institutional policy if an internal media library is used.

Slide presentations are suitable for delivering lectures to small and large groups. Available software makes it easy to create, update, and incorporate verbal and visual contents at minimal cost. The file is readily transmissible and portable. When presented by an unskilled trainer, however, the slide show may become a canned presentation. You should use a slide presentation as an aid and not be dependent on it as your lecture notes. Improper use is worse than no use at all (Brier and Lebbin, 2009, 358). Other than overuse and misuse, there are disadvantages. Projection equipment is needed, which diminishes the portability aspect. Although the size of projectors is becoming smaller, the resolution and projection quality of larger projectors are generally higher. If audio is played through a computer, external speakers may be needed. Projection and audio equipment for large groups is expensive.

6.2.3 Video/Audio

Video and audio recordings can be inserted into a slide presentation, or they can be played separately. They may be created in-house or purchased. These materials are helpful complements to various instructional strategies. Properly selected music, for example, can be used to set the right mood in collaborative learning. Video modeling has been found to be superior to tutorial training in computer skills even when both methods provide the same content and opportunity for hands-on practice (Gist et al., 1988, 262–263). A video is excellent for showing a dramatic event or action, which compels attention and is remembered (Dale, 1946, 103). A video can enlarge items that are normally difficult to see, such as components of an integrated circuit board. Using prerecorded video or audio material allows for standardization—different groups of trainees will always see the same demonstration. Another advantage is that it offers a common experience to a linguistically diverse group. Although different cultural groups may interpret some visual symbols differently, as explained in Chapter 16, most would have a common understanding of the basic ideas in a story. A video can be used to demonstrate performance of a manual task in psychomotor skills training. DeAmicis (1997, 157) finds that a video can be easier for trainees to visualize and understand than a live demonstration. Furthermore, the material can be replayed many times as needed. With instantaneous recording and playback, trainees' performance can be recorded for review and feedback. That is extremely helpful in validating competence or identifying areas for improvement. It is particularly applicable if the time required to perform a task efficiently is crucial, as in first-aid training, because the video playback shows exactly how many seconds or minutes have elapsed during task performance.

Whether the video or audio recordings are part of a slide presentation, equipment is needed to play them. The cost of video production, if done professionally, can be high. If a ready-to-use video package is purchased, the cost may also be high when the price of the package is based on the number of users. Additionally, the contents must be screened for accuracy and appropriateness. As an example, commercial safety training videos have been found to show individuals not wearing the proper personal protective equipment. Besides, you should be familiar with the material before showing it to your trainees. As in the case of media used in slide presentations, copyright ownership must be respected.

6.2.4 Easel Pad, Dry Erase Board, Electronic Copyboard

An easel pad may seem mundane compared with a fancy multimedia presentation, but it has its place in training. It is versatile. You can write information on it before the training begins to draw trainees' attention to certain points. You can use it during guided discussion to capture ideas that come up spontaneously. If a breakout session follows a general discussion, the notes taken on pages of the easel pad during the general discussion can be separated and given to the subgroups for reference during their breakout. Since easel pads are inexpensive, each subgroup can be given an easel pad to write down its discussion points for later presentation to the whole group. Alternatively, pages of easel pads with self-stick adhesive can be divided among subgroups easily. They are convenient for posting on the walls for review.

Dry erase boards serve similar purposes and are reusable. They are more expensive and not as portable, but less so compared to electronic copyboards. However, the advantage of the electronic copyboard is the built-in scanner that scans and saves what is written or drawn on the board. This is a considerable benefit because the saved file can be distributed or shared effortlessly.

You or whoever is leading a discussion may have to assign a scribe as it is hard to write while speaking. The writing needs to be legible. Unless an electronic copyboard is used, transferring the written information to an electronic document requires transcription and takes time. Even the larger electronic copyboards and dry erase boards are too small for a large group to view, and as the size increases, the board is less portable.

6.2.5 Model, Prop

This category includes actual equipment, replicas, miniatures, or other items that help to demonstrate an idea or task or to create a certain effect like a stage prop. For instance, firefighting training may use props that include fire extinguishers and a barrel of flammable liquid set on fire. The use of a model or prop activates the senses of sight and touch and, in some cases similar to this firefighting training, other senses such as smell may be involved.

Models and props are excellent for demonstrations and practice or role-playing, especially when the learning objectives are in the psychomotor domain. These training aids provide trainees with a realistic, hands-on experience. Trainees can see and feel the object. The learning process involves multiple senses, an advantage according to Dale's (1946, 42) "cone of experience."

Careful preparation to get the model or prop ready for training may be time-consuming. Safe use, transportation, and storage may be a challenge. Making a replica or miniature may incur time, cost, or skills that a trainer does not have.

6.2.6 Costume

Not used as much as its potential warrants, costume can work in at least two ways. It can identify a profession, which can help to build rapport. If your trainees are a homogeneous group in a profession that wears certain attire or uniform, and you are an "outsider," you can put on the same outfit. This promotes acceptance among the trainees that you are part of their group and breaks the "we" and "they" barrier. The uniform may be a simple T-shirt or polo shirt with a logo or slogan of the group. For example, a registered dietician training an athletic team on nutrition may wear the team's T-shirt. Similar to theatrical performance, costume can be used in training to portray a character, as when a trainer wears a costume to play a skit to emphasize a point. If done skillfully, a dramatic performance is memorable and conducive to learning retention. Accompanying props may be needed. One trainer showed up in gypsy costume, carrying with her a crystal ball, and played fortune-teller "predicting" success of the trainees' team.

When used properly, costume is a powerful tool that makes the training entertaining and impressive. A skit can be an icebreaker, or it can be used as part of a demonstration. Some costumes are expensive, although with creativity you can improvise with what you have in the wardrobe. To put on a good show, preparation and rehearsal are critical and take up a lot of time. A major drawback, and perhaps the reason why costume and performance are not widely adopted in training, is that the trainer must feel comfortable acting.

6.2.7 GAME

Many games require players to explore scenarios, make decisions, and solve problems. They are also entertaining. By making learning a fun experience, training games are effective as a training aid. They are found free or for purchase on the Internet, or you can create them from scratch or from templates. There are icebreaker games, design games, team-building games, motivational games, brainteaser games, and many others, designed for various group sizes and durations. Some of them are based on popular television game shows, such as *Jeopardy* and *Who Wants to Be a Millionaire*, or games that most people have played since childhood, such as tic-tac-toe. Familiarity makes it easy for the trainees to understand how to play the game and enjoy it. While board games are available, in this electronic age, a lot of the games are computer based and designed to be shown on a screen. They promote self-confidence, skill transfer, and knowledge application (Connolly and Stansfield, 2006, 473). They engage trainees and avoid boredom. Gaming, with the added uncertainty of playing for stakes, has been found to increase the affective response in learning tasks, which suggests enhanced engagement with the learning (Howard-Jones and Demetriou, 2009, 533–534).

Games may be played at the beginning as an icebreaker or throughout the training event. An icebreaker activity helps trainees relax and get to know one another. Games can be added spontaneously to stimulate participation or foster team collaboration. For instance, at some point during the training event, you feel that some trainees need more motivation to actively participate. You can divide the group into several competing teams to play the game of *Jeopardy*, with questions related to the training materials and prepared in advance. Games can be used as an assessment tool to check for understanding during or toward the end of the training.

There are also disadvantages. Games suitable for the group size and purpose must be carefully selected. A game should not be played just for fun if it does not improve the learning experience. Even well-chosen games may not appeal to everyone in the group due to cultural backgrounds or personal characteristics. You must be able to maintain control of the time. If the game is competitive, you must ensure fairness in the competition.

6.2.8 COMPUTER, INTERNET, SIMULATOR

Computers have been used for a long time by trainers for multimedia presentation of materials or by trainees for practice, self-guided discovery, or collaborative learning. The availability of high-speed Internet access in practically any training venue has made it convenient to obtain web-based supplemental materials on the spot. A hyperlink in a slide presentation can bring up the desired website and information without having to download the material prior to the training, which may make the file size of the presentation unmanageable. Trainees can search the Internet for information related to case studies or other participant activities.

More advanced use of computers and the Internet in training are found in computer-based training (CBT), e-learning, mobile learning, and other kinds of virtual training. Computers provide hands-on experience when training computer-related tasks and can serve many purposes in other types of training. CBT reduces instruction time compared with traditional methods (Kulik and Kulik, 1991, 90–91). With proper design it can incorporate activities that enhance the learning experience. CBT can be stand-alone or complementary to in-person training that is used in blended training. Tracking completion and test results is easy with CBT. Compared with classroom training, CBT may produce cost savings. The advantages of using technology cannot be realized if some trainees are not computer literate. The need for access to computers may present a problem in field training. Setup costs can be high. CBT is not effective for all training topics. Welsh et al. (2003) note that, in the training of psychomotor skills, e-learning is better for the declarative elements (e.g., the names of instruments) than the procedural elements (e.g., how to use the instruments). For some topics, CBT may be suitable only for refresher courses.

Simulators can be relatively simple software applications that perform calculations based on data input or complicated devices such as flight simulators that resemble the actual equipment and programmable for varying operating conditions. They can demonstrate a process at a speed faster or slower than the actual speed to illustrate the respective effects. Text, numerical values, or complex systems can be displayed in a form that is easier to understand. Simulation provides an opportunity for experiential learning and encourages trainees to apply theory to practice (Cleave-Hogg and Morgan, 2002). Novices can recognize and solve problems in a hazardous situation without endangering themselves or others (Tichon, 2007, 287). When trainees are health care providers, for example, simulation eliminates the risk of having inexperienced trainees practice on real patients. Experienced professionals such as Formula One race car drivers also use simulators since the amount of time they are allowed to practice on the track is restricted (BBC, 2013). Chapter 20 discusses virtual training in more detail.

6.3 CONSIDERATIONS FOR CHOOSING THE "BEST" AIDS AND MEDIA

While all aids and media may add impact to training, not all materials are suitable for all occasions. Using multiple types of media adds variety and allows for a change of pace. Using too much media may cause confusion and a loss of focus on the purpose of the training. You must decide "when" and "what."

The first thing to remember is that "teaching aids must be used in *integrated fashion*" (Dale, 1946, 494). Each piece of material has a purpose within the training plan. For example, in a training of emergency responders, a handout and slide presentation orient the trainees to the criticality of emergency response to saving lives, reinforced by a video that dramatizes the aftermath of a hurricane. Using models and props, the trainer demonstrates the roles of different types of emergency response personnel, supplies and equipment they need, and how to use the *matériel*. The trainees then practice with the equipment. Afterwards the trainer leads a discussion on problems that may arise in various scenarios and how to solve them. The resolutions are recorded on an electronic copyboard and distributed to the trainees later. The sequential use of the aids and media follows the logical flow of moving from knowledge-level learning to application and problem solving.

With regard to the "what," the selection should match the learning objectives, target audience, and trainer's skills, and adapt to situational constraints, as discussed in the following sections.

6.3.1 Learning Objectives

Learning objectives always constitute the most important consideration in selecting any strategy or material. The training aids and media must contribute to the content of the topic taught. As learning objectives move from the knowledge level to higher levels, you may change the media. Handouts have become almost indispensable. They provide reference materials and are suitable for knowledge-level learning objectives. They can accompany a lecture with multimedia slide presentation. Moving to the application level, easel pads can be used in a group discussion and trainees can practice in the field, with a model, or with simulation. Take the case of a training course on how to perform air quality assessment. The handouts can be a version of the slide presentation along with the laws and regulations. The trainees discuss how these would be applied to their individual practical situations and examples are noted on an easel pad. Having learned the concepts and methods of air quality assessment, they first practice using instruments to measure parameters such as emissions, climatic conditions, and other environmental factors, and then practice using a computer program that helps assessors determine the amount of air pollution based on the measured variables. This sequence would incorporate the use of multiple media to support multiple levels of learning objectives.

6.3.2 Target Audience

Characteristics of the trainees also occupy a prominent place in the training plan. If some trainees are not proficient in the language used in the training, it does not help to give them a stack of handouts in that language. These trainees may also find it hard to read other people's handwriting on the board. Computer literacy of the group is something to keep in mind. Individuals who are reserved may be reluctant to participate in games and activities. Dramatic actions and performances tend to have a universal appeal, even though it is still important to consider cultural nuances. Personal attributes are often not discovered before the training. That is why flexibility and skill on the part of the trainer are desirable.

6.3.3 Trainer's Skills

In a time management training course, a trainer wants to juggle three balls as a metaphor of people juggling their schedule between work, school, and family. That is a good idea, but the trainer has poor coordination and cannot catch a single ball. Should she play the juggler? Definitely not—unless she intends to emphasize the disastrous results of poor juggling of time demands. The saying "practice makes perfect" is tried and true. As in the case of selecting strategies, incidental trainers should begin with choosing aids and media that they find easy to employ, and practice using others until they feel comfortable.

6.3.4 SITUATIONAL CONSTRAINTS

Constraints are mainly related to group size, cost, and logistics. Certain materials such as easel pads are better used for small groups than large groups. Although slides can be presented on different types and sizes of display devices, a big screen and projection equipment are necessary for a large group. Group size is also important in the choice of games as it affects the time required and the dynamics. Some media may be too expensive to acquire. The perfect aid may not be available or practical. Ideally, trainees should see and practice with the "real thing" but how much resource will it take and how safe will it be for pilots in training to fly a plane the first time they get into the pilot's seat? That is why flight simulators are used instead.

6.3.5 SUMMARY

Table 6.1 presents a quick reference of noncomprehensive and nonexclusive training aids and media applications. Mix and match for variety. Be realistic about costs and logistics. Murphy's Law has no mercy on incidental trainers. Have a "Plan B" in case technology or equipment fails. For instance, you should be prepared to deliver the contents of your slide presentation without the slide projection when you find out at the training venue that the only projector has broken!

Finally, and most importantly, keep the focus on the learning objectives and use the media for reinforcement only—they are just the means to an end.

TABLE 6.1
Noncomprehensive and Nonexclusive List of Training Aids and Media Applications

Training Aid or Medium	Major Application
Handout	Reference
Slide presentation	Lecture, demonstration, action
Video/audio	Demonstration and practice, action
Easel pad/dry erase board/electronic copyboard	Group discussion
Model, prop	Demonstration and practice, role-playing
Costume	Demonstration, skit
Game	Group participation, assessment
Computer, Internet, simulator	Lecture, demonstration and practice, self-guided discovery, collaborative learning, assessment, refresher course

REFERENCES

BBC. 2013. "Trying out Fernando Alonso's Ferrari F1 simulator." *BBC News*, April 19. http://www.bbc.co.uk/news/technology-22218163.

Brier, D. J., and V. K. Lebbin. 2009. "Perception and use of PowerPoint at library instruction conferences." *Reference and User Services Quarterly* 48 (4): 352–361.

Cleave-Hogg, D., and P. J. Morgan. 2002. "Experiential learning in an anaesthesia simulation centre: Analysis of students' comments." *Medical Teacher* 24 (1): 23–26.

Connolly, T., and M. Stansfield. 2006. "Using games-based eLearning technologies in overcoming difficulties in teaching information systems." *Journal of Information Technology Education* 5: 459–476.

Dale, E. 1946. *Audio-Visual Methods in Teaching*. New York: Dryden Press.

DeAmicis, P. A. 1997. "Interactive videodisc instruction is an alternative method for learning and performing a critical nursing skill." *Computers in Nursing* 15 (3): 107–160.

Durso, F. T., V. L. Pop, J. S. Burnett, and E. J. Stearman. 2011. "Evidence-based human factors guidelines for PowerPoint presentations." *Ergonomics in Design: The Quarterly of Human Factors Applications* 19 (3): 4–8. doi:10.1177/1064804611416583.

Gist, M., B. Rosen, and C. Schwoerer. 1988. "The influence of training method and trainee age on the acquisition of computer skills." *Personnel Psychology* 41 (2): 255–265.

Howard-Jones, P. A., and S. Demetriou. 2009. "Uncertainty and engagement with learning games." *Instructional Science* 37 (6): 519–536. doi:10.1007/s11251-008-9073-6.

Kulik, C.-L. C., and J. A. Kulik. 1991. "Effectiveness of computer-based instruction: An updated analysis." *Computers in Human Behavior* 7: 75–94. doi:10.1016/0747-5632(91)90030-5.

Szul, L. F., and D. E. Woodland. 1998. "Does the right software a great designer make?" *T H E Journal* 25 (7): 48.

Tichon, J. 2007. "Training cognitive skills in virtual reality: Measuring performance." *Cyber Psychology and Behavior* 10 (2): 286–289. doi:10.1089/cpb.2006.9957.

Welsh, E. T., C. R. Wanberg, K. G. Brown, and M. J. Simmering. 2003. "E-learning: Emerging uses, empirical results and future directions." *International Journal of Training and Development* 7 (4): 245–258.

7 Physical Environment

7.1 TURNING (AN ALMOST) PERFECT TRAINING PLAN INTO A BOMB!

A registered environmental assessor in California was conducting onsite training for eight new members of his firm. He had everything planned out. The previous week he delivered the initial training in the classroom. For the onsite training, they were to practice how to conduct an environmental assessment of a piece of land intended to be purchased by one of the firm's clients for construction of an industrial complex. He already performed the "real" assessment and had all the handouts, including history of the site, from the project for illustration. Everything sounded like a perfect plan.

It was a sunny day in the fall with moderate temperature. The group carpooled in two vehicles and arrived at the site. The first hour went quite well and then, all of a sudden, the weather changed! The temperature dropped precipitously, and the wind speed picked up from moderate breeze to gale force. Unbeknown to the assessor and the group, inclement weather was forecast for the rest of the day. In the freezing cold and with the papers and drawings blown all over the place, the assessor had to terminate the training.

This example illustrates how failure to check one seemingly small detail on the physical environment—the weather forecast in this case—can destroy what otherwise would be a perfect training plan. And, it is not only at an outdoor location that the physical environment matters. As a trainer, you must heed the indoor environment as well.

7.2 KNOWING THE BASICS OF A SUITABLE PHYSICAL ENVIRONMENT

The physical environment is composed of everything seen, touched, or used during the training, from room layout to illumination to noise level and more. It is an element in the training plan that must be designed according to the learning objectives, instructional strategies, and media. For instance, if a large piece of equipment must be brought into a training room for the purpose of demonstration and practice. the room should have plenty of space for the equipment as well as for the trainees to move around. The physical environment should be arranged or modified based on special requirements of the training and the audience. Basic principles include the following:

- The physical environment should facilitate learning.
- It should be easily accessible by all trainees.

- It should allow trainees to see and hear all the training and participate in the activities.
- It should be safe, especially if demonstration and practice are involved.

Considerations of the physical environment are as pertinent to in-person training as to distance learning. The latter presents more situations that a trainer is unable to control. As the trainer, you should aim at managing what you can control and handling what you cannot.

7.3 MANAGING WHAT YOU CAN CONTROL

Just as employees are more productive working in a comfortable physical environment (Carnevale and Rios, 1995, 222), trainees are more attentive and learn better when situated in a comfortable environment. The training topic may dictate whether the training is conducted indoors or outdoors. The environmental factors to be considered are common in both cases, although implementation of strategies may be different. These factors are

- Advance site inspection
- Room layout and seating
- Lighting and noise
- Climatic conditions
- Water and sanitation
- Safety and evacuation routes
- Equipment and supplies
- Other logistics

7.3.1 Advance Site Inspection

Whenever possible, check out the training site prior to the training event, whether it is a corporate training room, a conference center, or a field location. This site inspection will be an opportunity to consider any necessary adjustments in the training plan with respect to the physical setting and room or site layout. For example, you want to eliminate distractions such as interesting views from any window of a training room. If you find that the room has windows looking through the hallway or another room where people may pass through, carry on conversations, or perform activities, you can prepare posters of a suitable size to cover the windows. The posters can be the motivational type or relevant to the topic of the training. Having said that, windows that provide a nice view, say, of a long beach with white sand and blue water, can offer relaxation during breaks, so you may want your posters to be easily removable during a break and reapplied afterwards.

If a visit to the site is not feasible, find out as much as possible about the venue. Facility services or other contact persons should be able to provide information on corporate training rooms. Conference centers and hotels publish floor plans on their websites, or their sales department would have the information. Normally a trainer who performs training at a field location would be familiar with the site or have

Physical Environment

a local contact that is knowledgeable. Maps and street views found on the Internet can be helpful.

Outdoor training is usually planned for activities, and seating is not needed. Other than the basic requirements, the main consideration is that space is sufficient for the activities. For indoor training, room layout and seating should be well thought out.

7.3.2 Room Layout and Seating

A fundamental of room layout is that all trainees should be able to see the trainer and all training aids and media used. They should also be able to participate in the training activities. The most common room layouts for training are theater, rounds, classroom, H-square, U-shape, and conference, as depicted in Figure 7.1. As a general rule, leave plenty of aisle space between tables, and between the walls and the seats closest to the walls. Theater, rounds, and classroom styles can accommodate larger groups than the other setups. Theater style is more conducive to lectures than team activities. A drawback is that trainees have no table that would make note-taking easier, whether by hand or on their electronic devices. Rounds facilitate interaction among trainees. If rounds are used, be sure to arrange the chairs in such a way that no trainee would have his or her back to the projection screen, easel pad, or white board if any of these training aids is used. Rounds and classroom, and sometimes theater, layouts allow for easy passage in and out of a seat and are preferred when there are activities that require physical movement. H-square, U-shape, and conference layouts limit the size of the group; however, they are desirable for small group discussions and activities. In some situations, it may be appropriate to change the room layout for different portions of the training event. For example, an introductory lecture can take place in a large classroom, followed by breakout sessions in conference rooms.

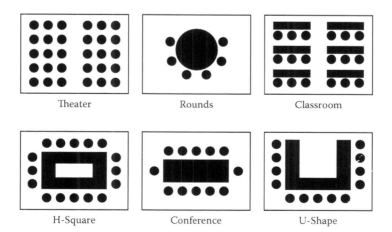

FIGURE 7.1 Common room layouts for training.

The room layout is affected by other requirements. If trainees will be using their own computers or tablets, power outlets must be close by where they sit or extension cords must be run. If all trainees are to use computers or other equipment provided to them, the training will likely take place in a room specially designed for the purpose.

Comfortable seating is vital for trainees to be able to concentrate on learning. A school classroom with desks and chairs for kids is not appropriate for training adults. Hotels often must be instructed as to the size of the room needed and the number of tables and chairs required, in addition to how these should be set up, or they may put too many people in a room or at a table, or leave a center aisle only in a classroom without aisles on the sides. The floor plans and seating capacities available from hotels do not always serve a trainer's purpose. A rule of thumb is to allow 3 ft of table space per person. Naturally, this requirement may change depending on the trainee population.

An electronic or paper training announcement or registration form should always ask if an individual has special needs that require physical accommodation. A person in a wheelchair is often seated in the front or in the back—in the front to facilitate participation in activities and in the back for ease of ingress and egress. Make the appropriate arrangement based on the training plan.

7.3.3 Lighting and Noise

You want the light intensity to be right for the kind of task being taught, the training media being employed, or the activities being planned. Avoid shadows or glare, which can come from a light source or from reflection (Chengalur et al., 2004, 567, 569–570). This is especially important if the training requires much use of computers by trainees. In such a case, it is preferable to conduct the training in a computer lab, although the choice is also dependent on other factors. Use high-quality projection equipment and good slide design for slide presentations so that it is unnecessary to dim the room light, particularly if the time of the day is immediately after lunch. A dark room after lunch is perfect for a siesta!

Noise is unwanted sound that may interfere with communications and cause distractions (Chengalur et al., 2004, 578). Loud background noise may be inevitable in a field setting, for example, when the training takes place at a construction site. In that situation, try to find a quiet location nearby to explain the procedures before bringing the trainees into the noisy environment. Alternatively, use special communication devices such as noise-canceling radio headsets with earmuffs that allow clear communication and provide hearing protection from overexposure to noise. In training rooms, eliminate noise in the surroundings to the extent possible. If the training event takes place in a hotel meeting room, do not allow the hotel to schedule in an adjacent room another group that may create a lot of noise, such as a band.

7.3.4 Climatic Conditions

"I was freezing in that room!"
"It's so warm in there I felt I couldn't breathe!"

Physical Environment

Room temperature is one of the most common complaints about the physical environment. It is hard to concentrate when the temperature is too cold or too warm. It is also hard to please everybody because individual preferences vary. The comfort zone for 80% of building occupants is when the temperature is between 20°C (68°F) and 23°C (74°F) and the relative humidity is about 50%. (This is an approximation from the guidelines of ASHRAE [2010, 6], which require detailed computations based on clothing, air speed, and other factors.) The temperature range in the comfort zone increases as the relative humidity decreases, and vice versa.

The number of people in a room and the amount of physical activity affect thermal comfort. When these factors change substantially, the room temperature may need adjustment. Outdoor climate cannot be controlled, but it is wise to consult the weather service and schedule training on days when the weather is mild, to the extent possible.

7.3.5 Water and Sanitation

Unless the training is meant to be a physical endurance boot camp, the availability of water and restroom facilities is imperative at any site. Drinking water should be provided even if other beverages are supplied. This is especially critical in heat stress conditions because drinking water, not carbonated soft drinks, prevents dehydration. Experience from meeting planning also shows that on the average attendees consume equal amounts of bottled water and all other cold beverages combined. Breaks should be scheduled about every hour, more often if many physical activities are performed in hot or cold weather.

7.3.6 Safety and Evacuation Routes

Even if it is a boot camp, the safety of trainers and trainees should be a high priority. This is most important for field training that takes place outdoors. Young, physically fit recruits have died of heat stress during military training. With an aging workforce, you must be cognizant of some trainees' special needs. The indoor environment is not free from hazards either. Emergencies such as a fire may occur and you must be prepared to deal with them, including evacuating trainees who may need assistance. This is another reason why trainees should be encouraged to inform you beforehand about any need for special accommodation. During the introduction at the beginning of the training, point out emergency exits and evacuation routes. The trainees would appreciate the fact that you care about their safety.

If the nature of the training itself can incur safety hazards, as when teaching how to handle a large chemical spill, there should be enough trainers or training assistants to oversee the safety of the trainees. For most field training, the ratio of trainees to trainer should not exceed 10:1 (National Environmental Safety and Health Training Association, 2004, 35). The ratio should be lower if the probability of injury is high.

Advance inspection of the training site gives you an opportunity to anticipate and mitigate hazards. Trainees would appreciate advance notification of physical demands and potential hazards, along with directions in how to protect themselves, such as the types of clothing and shoes that should be worn. Without warning, trainees may be oblivious to precautions that would seem fundamental to you, for example, wearing nonskid shoes for climbing up the steep hillside of a landfill. Inform trainees beforehand what they can expect and, immediately prior to beginning the field training, reemphasize the criticality of following safety instructions.

7.3.7 Equipment and Supplies

Before the training event, make sure that the training aids, media, and equipment on your training plan are in hand and check for proper functioning. If someone else is responsible for the logistics, be sure to coordinate with that person well ahead of time and follow up to ensure that everything you need will be there. Whenever possible, incorporate redundancy in your planning. For instance, if a model is critical for your demonstration, try to have two of the models ready. If you need a computer for a slide presentation, bring an extra laptop.

Again, safety is a priority. If using electrical outlets, have extension cords and gaffer's tape available. Gaffer's tapes are preferable to other adhesive tape as they leave no residue. At the training site, make sure that all equipment is placed on stable surfaces. Use the tape to secure extension cords to prevent trips and falls. Do not overload extension cords or power outlets.

Attention to detail shows professionalism and adds to your credibility as a trainer. A perfectly designed slide presentation lacks a professional look if part of the projection falls off the screen or when the slides are projected as a trapezoid instead of a rectangle, due to failure to properly set the zoom ring or the keystone adjustment of the data projector. Whether the site has facilities personnel or audiovisual technicians responsible for setting up equipment, it is your responsibility, as the trainer, to see that everything is in order.

7.3.8 Other Logistics

When the training will last for more than a couple of hours, consider serving refreshments to prevent loss of concentration due to hunger or thirst. Refreshments should be light, not substantial food, or the trainees may become drowsy. At least drinking water should be provided. Place the order for food and drinks early and, on the day of the training event, verify that the correct items and quantities are delivered. Arrive at the training site with plenty of time to set up equipment and supplies, confirm that everything is correct, and make last-minute adjustments as needed. These tasks should be completed before the first trainee arrives. Be prepared that some early birds may show up 30 minutes or more before the scheduled time.

When the training venue may be unfamiliar to some trainees or is hard to find, provide directions in advance. If referring trainees to obtain maps and directions from websites, check that such information is correct as sometimes they are not. You do not want trainees to arrive late and exhausted because they had difficulty finding the place.

Physical Environment

7.4 HANDLING WHAT YOU CANNOT CONTROL

Ideally, you will be able to set up the physical environment in exactly the way you want. It is not always possible, particularly with an outdoor venue or if the site is owned by a third party, such as a hotel. Two challenges that are beyond the trainer's control and occur a lot are distractions and poor furniture design. They are exacerbated in e-learning and other circumstances where the training has a self-study component. E-learning is a form of virtual training and will be discussed in Chapter 20. For now, let us think about issues that may arise during in-person training.

7.4.1 DISTRACTIONS

Other than a window view as mentioned in Section 7.3.1, a common source of distraction is noise. Field setting at a construction site has been mentioned. Even when training takes place inside, if the training room does not have good acoustic properties, noise from the vicinity, such as people's conversations, could be distracting. One suggestion is to politely request a "quiet zone." In the corporate setting, typically people are amenable to such a request. Put up signs to that effect in the surrounding area to remind those who are aware of your request and to alert others who happen to come by. At a venue owned by a third party, it is always prudent to include in the contract for the facility rental a condition that prohibits the facility to place a noisy group next to your meeting area.

Another distraction is thermal comfort, or the lack thereof. Besides the fact that thermal comfort varies much among individuals, you and your trainees may be outside, where you cannot control the climatic conditions. Steps to prevent heat- or cold-related disorders are crucial. At a minimum, if there is any chance of encountering inclement weather, make sure supplies and materials that may be necessary under those conditions are available and alert the trainees to be prepared accordingly. In some buildings, occupants are unable to set the thermostat and either no one from facility services is available or it takes a long time to fetch maintenance personnel. Having both hot and cold beverages available, including hot drinking water, is advisable to help those who feel cold to warm up and those who feel hot to cool down.

For distractions that these strategies cannot alleviate, you will have to use your presentation and facilitation skills to keep the trainees' attention. Chapter 9 and Part 2 have suggestions on presentation and facilitation.

7.4.2 FURNITURE DESIGN

The conference center of a large hotel has recently undergone renovation. All the interior furnishings of the meeting rooms are brand new. The tables for classroom seating are no longer the standard, rectangular 6-ft tables covered with a tablecloth. Instead, the tabletop has a gemlike shiny finish, in a decorative pattern of sapphire color, and reflects the light from the splendid crystal chandeliers. It is beautiful, even majestic! But, the reflection produces glare—not an ergonomically sound design for classroom training. If you are in a similar situation, tactfully request the conference center staff to add table coverings. Glare is a distraction and affects vision.

Not all conference centers or hotels try to impress meeting planners and attendees with fancy tables. What happens very often, however, is that the seats of the chairs are too hard or too soft and the chairs are not height-adjustable, making it uncomfortable to sit on them for an extended period of time. Auditorium seating may also cause discomfort, for reasons such as insufficient legroom. You cannot change the furniture. What you can do is to build more frequent stretch breaks into your training schedule. They do not need to be formal breaks on the agenda. When you come to a stopping point, have the trainees stand up and lead them, or ask a volunteer to lead them, in a stretching exercise. These microbreaks can take only 1 to 2 minutes. They benefit the body and mind (Fenety and Walker, 2002, 586).

7.5 PAUSING AND REFLECTING ON YOUR TRAINING PLAN

A training course rich in content and practical in application can be ruined by delivery in a physical environment that does not promote learning. Advance planning and site inspection can help you avoid potential problems; otherwise, the problems may affect achievement of the learning objectives and test results.

At this point, your training plan has incorporated most of the elements that you will need to deliver the training. You may want to take a quick review of the plan to ensure that the instructional strategies, aids and media, and physical environment are convergent with the learning objectives and are practical enough to be executable, before moving on to the next step, which is deciding on the assessment methods.

REFERENCES

ASHRAE. 2010. *ANSI/ASHRAE Standard 55-2010: Thermal Environmental Conditions for Human Occupancy*. Atlanta: ASHRAE.

Carnevale, D. G., and J. M. Rios. 1995. "How employees assess the quality of physical work settings." *Public Productivity and Management Review* 18 (3): 221–231.

Chengalur, S. N., S. H. Rodgers, and T. E. Bernard. 2004. *Kodak's Ergonomic Design for People at Work*. 2nd ed. Hoboken: John Wiley & Sons.

Fenety, A., and J. M. Walker. 2002. "Short-term effects of workstation exercises on musculoskeletal discomfort and postural changes in seated video display unit workers." *Physical Therapy* 82 (6): 578–589.

National Environmental Safety and Health Training Association (NESHTA). 2004. *Accepted EHS Training Practices: An Implementation Guide*. Phoenix: NESHTA.

8 Testing and Assessment

8.1 CRITICALITY OF TESTING AND ASSESSMENT

Think about training courses you have attended yourself as a trainee. Were you tested every time on what was taught? It is amazing how often training is completed without assessing whether the learning objectives have been met. Some say that as much as 85% of all training does not incorporate any form of testing.

The purpose of training is to transfer knowledge, teach skills, or change behavior or attitude. Without testing and assessment, how does a trainer know if this purpose is achieved? Testing is the means to verify that the trainees have indeed acquired the knowledge or can perform the task as intended. Testing may be used also to gauge whether trainees understand organizational values and goals meant to be instilled through training; however, the actual adoption of those values and goals into daily practice may be better revealed in a subsequent assessment using other methodology.

Testing results help a trainer assess if the instructional strategies and training delivery are effective. If the training has been conducted to comply with regulatory requirements or internal policies, testing is needed to document that the trainees are competent to perform the work.

As individuals, you probably have taken many tests in your life, from the days of elementary school to the time when you enter a profession. The bar exam that qualifies attorneys to practice is one example of numerous licensing and professional examinations. These examinations are important indicators to those within the profession and the general public that the individuals who have passed have achieved certain levels of competency in that profession. Do you know the bases of these tests and why they are designed the way they are? State licensing boards and professional certification bodies that are accredited by accrediting agencies in the United States—the American National Standards Institute (ANSI) (2011), the Council of Engineering & Scientific Specialty Boards (CESB) (2008), and the Institute for Credentialing Excellence (ICE) through its accrediting body National Commission for Certifying Agencies (NCCA) (2012)—have common principles of designing their examination questions. Some of these principles are discussed in Chapter 13. This is a topic you ought to understand as an incidental trainer. And, it might even help you score higher in your next licensing or professional examination!

8.2 APPROACHES TO TESTING

Before developing the actual test, decide how you want to conduct the testing. Should there be a pretest as well as a posttest? Will the testing be norm-referenced or criterion-referenced? Will you employ formative testing in addition to summative testing?

8.2.1 Pretest and Posttest

A pretest is sometimes used in conjunction with a posttest. As the name implies, a pretest is administered before the actual training begins. The pretest consists of the same questions or observations as the posttest that is administered after the training. The expectation is that a trainee will achieve a higher score in the posttest than the pretest—evidence of learning having taken place. A psychological benefit for the trainees is that an increase of posttest scores from the pretest scores may give them a sense of accomplishment and motivate them to want to learn more in future.

Trainers use the pretest and posttest combination as a convenient way to assess improvement of performance after training. However, most of the time only one group of trainees is involved. Nau et al. (2009, 205) evaluated a training course in managing patient aggression for nursing students using this method. The authors analyzed the confidence score of the students in handling patient aggression before and after the training course. It was found that the confidence score increased significantly after the training. Only one group of trainees was studied. To truly evaluate the effectiveness of training, there should be an experimental group that receives the training and a control group that does not receive the training. Comparison of the results of the posttest of the two groups will better reflect the change due to the training by minimizing the effects of confounding factors such as repetition. Unfortunately, this type of experimental design is rarely used in practice to determine the significance of training. Also, some authors advocate a pragmatic approach, without a pretest or experimental group, when the aim is to have trainees achieve a certain performance level rather than to quantify change. According to Sackett and Mullen (1993, 618), the distinction is that a change-oriented evaluation attempts to determine whether a training program is producing results as designed, whereas a performance-oriented evaluation aims at determining both the effectiveness of the training program and the competence of the trainees after completion of training.

Estimating change or performance is not the only reason why a pretest has efficacy. Another reason for a pretest is to allow the trainer to estimate the current knowledge of the trainees and modify the training content or delivery method if necessary. You will find this approach useful if you are not familiar with the particular group of trainees, or if you are aware that the current levels of competency among the trainees vary and you want to find out the extent of the variation in order to adapt the training to your audience. Continuing education courses, whether delivered in-house or by a third-party contractor, often have diverse attendees with varying years of experience and degrees of keeping up-to-date with current issues in their professions. This is a situation where a pretest may be helpful in deciding how much they already know about the subject matter of the training.

8.2.2 Norm- and Criterion-Referenced Testing

Norm-referenced testing is grading "on the curve." A trainee's test result is compared with the performance of other trainees and the trainee's "*relative*" standing along the continuum of attainment is the primary purpose of measurement" (Glaser, 1963, 520). An individual may attain a high score by performing better than the rest of the group, but even the person with the highest score may not meet the task performance requirement. As a consequence, this approach is not desirable for evaluating job performance, where generally a minimum level of competency is required for a specific task. Licensing boards and accredited professional certification bodies do not use this approach in their examinations. These examinations are criterion-referenced.

In contrast to norm-referenced testing, criterion-referenced testing requires that the trainee's performance be compared to a set standard that is required of an individual performing a job or practicing a profession. The score represents the degree of achievement compared with desired performance (Glaser, 1963, 519). A trainee must achieve a minimum score in order to be given a passing grade, without regard to other trainees' scores. The passing score, or cut score, is meant to translate into the minimum amount of knowledge or acceptable behavior the trainees should have; therefore, it is possible that no one passes at all. An example of a criterion would be answering 80% of a set of questions correctly. When this approach to testing is used, the passing score should be stated in the learning objectives or communicated to the trainees in some way.

How would you determine the passing score? Many trainers use 70% or 80% somewhat arbitrarily. In proper training and testing, the passing score should be established based on the work requirement. The passing score and its justification should be documented. For example, an inspector in an apparel factory may need only 80% accuracy performing quality control inspections of garments, but a much higher degree of accuracy and multiple checkpoints would be required in a pharmacy distribution center. The passing score in high-stakes tests, such as many licensing and certification examinations, is validated using psychometric techniques.

Note that the term *criterion* refers to "the behavior which defines each point along the achievement continuum" (Glaser, 1963, 519); thus, different criteria can be used at several stages of the training before the end of the course. In other words, they can be applied to formative testing as well as summative testing.

8.2.3 Formative and Summative Testing

As you progress through the agenda of a training course, how do you know if the trainees are "getting it"? A good way to find out is by using formative tests. They do not have to be done in a formal way. They are exercises that test trainees' level of competency but do not count in the final grade. The purpose is to enable you and your trainees to make decisions on the spot about the progress so as to improve the learning process or outcome (Black and Wiliam, 2009, 9). Clearly, most of the time you would be the one initiating any adjustment and implementing corrective instructional strategies in view of unsatisfactory formative test results. For example, during a training course on the use of website design software, the trainer may demonstrate the steps of putting a simple web page together, with text and a few hyperlinks. He may have the trainees follow his steps and do each step after him.

He then may ask the trainees to individually design a web page of their own to make sure that they understand and remember how to do it. At this point, if it appears that all trainees are capable of working on their small project and producing web pages that are acceptable, he can move on to the next stage and teach them, in the same manner, the design of a more complex web page with multimedia links and animation. On the other hand, if it turns out that most of the class has created web pages that have problems such as broken links, he may reassess the learning process and, in addition to the demonstration and practice, write down for the trainees each step for properly inserting links.

Formative tests are also a useful review tool for reminding trainees of the important points of one segment of the training before proceeding to the next. The review makes it easier for trainees to grasp the upcoming, new material.

A summative test is usually administered at the end of a training course and graded to evaluate competency. For a longer training course that lasts more than a day, a summative test may be administered at the end of several time intervals, such as a day or a week, and the final score will be a combined total or an average of these test scores. The average may be a weighted average if different portions of the training are considered not of equal importance.

8.3 RELIABILITY AND VALIDITY

Test items are meaningless unless they are both reliable and valid. What does that entail?

Reliability necessitates that the measure yields consistent results when applied to different trainees with the same level of ability. In other words, a reliable test discriminates those who have mastered the knowledge or skill from those who have not. If two trainees in the same course have achieved the same level of competency, they would achieve similar scores in a test. Their scores would be quite different from those of another trainee who has not acquired the skills taught.

A consistent result, however, is necessary but not sufficient (Pedhazur and Schmelkin, 1991, 81). The test metric must also be valid, which means that it measures the desired outcome. Suppose the objective of a training course for a group of call center agents is learning how to improve customer satisfaction. What would be a valid metric of performance improvement? Would it be more calls handled per agent per day? All the "star" agents may be handling 10% more calls—consistent and reliable test results—but there may be no change in customer satisfaction. Improved scores in a customer satisfaction survey are a valid test measure, whereas an increased number of calls handled per agent per day is not (although expeditious resolution of customer concerns and reduced call duration may come into play in customer satisfaction). On the other hand, if the learning objective of the course is to increase the calls each agent can handle, by becoming more proficient with the software application used, the latter test measure may be valid. In the context of testing in training, a test is valid when it is reliable and criterion-referenced, and the criteria must be based on the learning objectives.

To ensure reliability and validity in the assessment, a number of factors must be considered. The two that are most fundamental are what should be measured and how you measure it.

What should be measured? At first sight, the answer seems straightforward. If the learning objective is at the knowledge level, you measure trainees' knowledge after the training; if the learning objective is to change behavior, you assess how the task is performed after the training. Knowledge is usually quite easy to measure, but performance under actual conditions is not always measurable. A new firefighter may be trained under all kinds of simulated physical conditions of fires, but the psychological impact would not be the same while fighting a real fire, compared to fighting a fire during the training and knowing that the training environment is controlled and relatively safe. One benefit of on-the-job training is that the trainee's performance can be observed in real conditions. In many other circumstances, the best measure possible may be performance under simulated conditions, which is a surrogate measure. With a surrogate measure, care should be taken that it correlates with the desired outcome. In other words, will the scores predict actual performance, since the ultimate goal of training is to enhance performance? If not, the testing method may not be valid. Chapter 13 discusses ways to improve reliability and validity of tests. The following section describes the most commonly used testing methods.

8.4 TESTING METHODS

The earlier example of the call center training brings home the fundamental point that you should match testing methods with the learning objectives. For instance, it would be inappropriate to assess competency in a psychomotor skill with a written test alone (Wan and Terrebonne, 2007).

Many testing methods are available and most can be administered in person or online. Table 8.1 summarizes types of testing methods applicable to different levels of learning objectives.

TABLE 8.1
Matching Testing Methods with Learning Objectives

Learning Objective Level	Testing Method
Knowledge	Recognition-type test items
Comprehension	Recall-type test items, role-playing
Application	Case study, performance, role-playing
Problem-solving	Case study, simulation, role-playing
Psychomotor	Case study, performance, role-playing
Affective	Case study, role-playing, observation, self or supervisor report

Note: The list of testing methods or their appropriateness for a particular level of learning objective is not exclusive.

8.4.1 Multiple Choice, Multiple Select, True/False, Matching, or Ordering

These tests are suitable primarily for testing knowledge-level learning objectives. Although the effort required of the test-taker may vary slightly, depending on how the questions are framed, these tests have one thing in common: The question itself incorporates the answer. For this reason, a person taking the test only needs to recognize the correct answer or make a good guess—or roll a dice! That may be why this type of question is popular with trainees. If there are four or five answer choices, a test-taker has a 25% or 20% chance of picking the right answer. Trainers and professional certification bodies also like these questions because they are easy to grade, either manually or electronically, for large groups. The knowledge-level testing is acceptable because certificants would be required also to provide proof of practical experience in the profession.

Sometimes such test items are used for problem-solving objectives. Instead of giving trainees one statement and having them pick one answer from multiple options, a scenario and a problem are described and one or more questions are written based on the scenario and problem. Typically, an answer key is developed by subject-matter experts or experienced employees familiar with that type of situation (McDaniel and Nguyen, 2001, 106). This is one form of situational judgment testing described in Section 8.4.4. A modified true/false question is another alternative. If a trainee marks a statement as false, he or she is required to write a statement that is true. This type of question is akin to the short-answer question.

8.4.2 Fill-in-the-Blank or Short Answer

These test items are useful for verifying comprehension. Trainees need to recall rather than merely recognize the correct answer. The question or the statement with the blank should be unambiguous and lead an informed trainee to the correct answer. A short answer comprising just a word or two can be graded automatically in a computerized test. The software would match answers to a preloaded list of acceptable answers. If a short answer requires more than a couple of words, one would expect more variations in the answers given by trainees. This aspect makes it harder or impossible to correctly grade the answers automatically. When the answers are short, it would not be too time-consuming to grade answers of a typical group of 25 or fewer trainees. Whether the grading is computerized or manual, always use a standard answer key as a yardstick to minimize subjectivity and ensure that the answers are to the point.

8.4.3 Essay or Oral Explanation

This method also requires trainees to recall the correct answer. Additionally, a trainee must have a good understanding of the subject matter in order to explain it in his or her own words. It can be used at the comprehension or higher levels. Again, it is not conducive to automatic grading. To maintain objectivity, grade answers against a list of points that should be covered in the answers.

Testing and Assessment 73

Another consideration is that the test result of a question that requires an essay or oral explanation may be affected by a trainee's writing or speaking skills, so it is biased against those who do not have good communication skills or whose native language is different from the language used in the training. Of course, if the training is on communication skills, the assessment would be equitable.

8.4.4 Case Study or Situational Judgment Testing

For learning objectives at the application, problem-solving, or affective level, a case study or scenario can be given to the trainees in a written or oral format. The trainees are asked how they would handle or feel about the situation. Oftentimes, there is not one right answer. The purpose is not to elicit one correct response from the trainees but to ascertain their understanding of the topic and ability to apply knowledge in a real-life situation. For instance, in a risk management training course, trainees may be given a description of a new program in an organization, such as an internship program, and asked to assess potential liabilities that the organization might be exposed to as a result of implementing the new program.

This testing method is suitable for formative tests as well as summative tests. Used as a formative test, it gives the trainer an opportunity to evaluate the progress of the trainees in their ability to apply what they are learning. It can also be used as a group exercise by presenting the scenario for group discussion. When used as a summative test, it is advisable to grade against a list of key points that are relevant to the situation, even when more than one point of view are expected.

8.4.5 Performance or Simulation

Performance testing is best for application or psychomotor skill learning objectives. The performance of the trainee is observed and compared to a standard checklist. The observation may occur during the training as a formative test or upon completion of the training as a summative test. Sometimes additional follow-up takes place a period of time after the training to verify the long-term result.

Actual performance is not always feasible or safe, especially when a trainee is a novice. With technology, simulation is an alternative. For example, flight simulators for pilot training and testing have been used for decades.

Performance or simulation may be used for testing problem-solving skills and are more realistic than using a case study or scenario. Subsequent to the simulation, testing in the actual work environment is still necessary—imagine yourself as a passenger in an airplane flown by a pilot who only had passed a test "flying" a simulator!

8.4.6 Role-Playing

Role-playing is not only an instructional strategy but is also a versatile testing method that can be used for learning objectives ranging from comprehension to problem-solving and affective levels. It is good for testing psychomotor skills too. Most of the time a trainee participates in role-playing with one or more other trainees or the trainer, so it is more involved than a straight performance by one person.

When you use this testing method, make sure that clear directions are given to the trainees as to what they are supposed to do and how their performance will be evaluated. In particular, be specific about whether all trainees in the same role-playing session will be graded as a group or individually. This is something that requires prior careful thought since individual performance may be affected by the action or inaction of others in the group. As in the case of performance or simulation, a standard checklist should be used in observing and grading performance.

The downside of using role-playing as a testing method is possible confounding of test results. Be aware that some trainees may be uncomfortable with role-playing or not skilled at thinking on their feet, thus adversely affecting their performance.

8.4.7 Observation or Report

Instead of observing performance of a task, this method involves observation of behavior, attitude, or values. It is especially useful for learning objectives at the affective level, which is hard to measure during or immediately after training. Observation over time may provide more relevant feedback; therefore, the trainee is observed at a specific time after the training to determine whether there is in fact a change in behavior, attitude, or values in line with the learning objectives. Rather than observation, reporting by the trainees themselves or their supervisors may accomplish this purpose as well. Data can be collected through interviews or questionnaires.

8.5 OBJECTIVITY AND SUBJECTIVITY

It is obvious from the nature of these testing methods that some are more objective than others. For instance, those that have only one correct answer if the test item is properly designed, such as multiple choice, are objective. Qualitative assessments like observation of change in attitude are more subjective. Even when a preset checklist is used in performance or observation testing, it is possible that results may differ when the test is administered by different trainers or by the same trainer at different times, so you should attempt to be as consistent as possible in judging whether the performance or behavior meets your set criteria. Chapter 13 has more on how to maximize objectivity.

8.6 NEXT STEPS

You have selected test methods based on the learning objectives. You have taken every step to ensure reliability and validity of the assessment. Your training plan is almost complete. While you still need to develop the course evaluation, you can get ready to put on a good show—deliver the course!

REFERENCES

American National Standards Institute (ANSI). 2011. "Accreditation Policy for Personnel Certification Accreditation Program: Public Policy PCAC-PL-501." ANSI. Last modified September 16. https://www.ansica.org/wwwversion2/outside/ALLviewDoc.asp?dorID = 355&menuID = 2#doc15298.

Black, P., and D. Wiliam. 2009. "Developing the theory of formative assessment." *Educational Assessment, Evaluation and Accountability* 21 (1): 5–31.

Council of Engineering and Scientific Specialty Boards (CESB). 2008. "Accreditation Guidelines for Engineering and Related Specialty Certification Programs." CESB. Last modified January 1. http://cesb.org/accreditation-guidelines.html.

Glaser, R. 1963. "Instructional technology and the measurement of learning outcomes: Some questions." *American Psychologist* 18 (8): 519–521.

Institute for Credentialing Excellence (ICE). 2012. "NCCA Accreditation." ICE. Accessed December 31. http://www.credentialingexcellence.org/p/cm/ld/fid=86.

McDaniel, M. A., and N. T. Nguyen. 2001. "Situational judgment tests: A review of practice and constructs assessed." *International Journal of Selection and Assessment* 9 (1/2): 103–113.

Nau, J., T. Dassen, I. Needham, and R. Halfens. 2009. "The development and testing of a training course in aggression for nursing students: A pre- and post-test study." *Nurse Education Today* 29 (2): 196–207.

Pedhazur, E. J., and L. P. Schmelkin. 1991. *Measurement, Design, and Analysis: An Integrated Approach*. Hillsdale: Lawrence Erlbaum Associates.

Sackett, P. R., and E. J. Mullen. 1993. "Beyond formal experimental design: Towards an expanded view of the training evaluation process." *Personnel Psychology* 46 (3): 613–627.

Wan, M., and N. Terrebonne. 2007. "A Critical Component in Safe Patient Handling: Validating Competencies of Patient Care Staff." March. Poster presented at the Safe Patient Handling and Movement Conference, Lake Buena Vista, FL.

9 Presentation and Facilitation

9.1 A TALE OF TWO INCIDENTAL TRAINERS

Two colleagues, Derek and Johnnie (not their real names), were safety supervisors of a midsize corporation. Both of them had more than 20 years' experience in this field. Each had supervised and mentored other safety professionals. The mentoring had been done on the job, one-on-one, since the company's training department had responsibility for all types of training, including safety. When the economy declined, the company decided to eliminate the training department. The safety manager—Derek's and Johnnie's boss—informed them that moving forward they would have to provide safety training at all new employee orientations. These meetings were scheduled once a month, with about 30–50 new employees in attendance each time. Derek and Johnnie would take turns to give the safety training presentation at the meeting. They would become incidental trainers.

Derek was alarmed. His immediate response was "I can't speak in front of a group!" His boss tried to convince him that it would be easy, because he had experience training other team members and he could use the training materials the training department already prepared.

Johnnie volunteered to be the first to give the training in two weeks' time. He was articulate and had experience in public speaking as a hobby. While preparing for the safety orientation, he picked up tips from friends who were experienced trainers to enhance his facilitation skills. The feedback from the new employees who attended his session was positive.

Another month went by fast. It was Derek's turn to conduct the safety training. He had familiarized himself with the training plan and materials, but he was unable to overcome his anxiety about speaking to a group. On the day of the new employee orientation, he gathered up his courage to show up at the auditorium, but he was so fearful that everything he knew about safety in the past 20-plus years escaped him! He managed to mumble through the notes he inherited from the training department and "survived" the allocated time of 45 minutes. The following 15 minutes were for a quiz to test the new employees' knowledge on the topic. As he found out later, the scores were extraordinarily poor, probably because the employees could not hear half of what he muttered.

Before Derek had to worry about the next time, the company decided to downsize further. The safety department was to eliminate one supervisor and two technician positions. Both Derek and Johnnie had been with the company for about the same length of time and were excellent safety professionals, but since Johnnie was more adaptable in different roles and had better communication skills, the manager decided to let Derek go.

The story of Derek and Johnnie highlights two points. First, no matter how knowledgeable an incidental trainer is on the subject matter, good communication skills are essential to transfer the knowledge to the trainees. Second, a good trainer should possess both presentation and facilitation skills. Johnnie was smart enough to know that. Although he was a good speaker, he prepared himself to be also a good facilitator.

9.2 PRIOR PROPER PREPARATION

"Prior proper preparation prevents poor performance of the person putting on the presentation" (The Princeton Language Institute and Laskowski, 2001, 17, 124). Preparation is vital to any kind of presentation, whether it is a platform speech or training. The following discussions relate to in-person training.

In training, preparation includes everything from conducting the needs assessment to developing the lesson plan, selecting instructional strategies, organizing the course materials and media, arranging the physical environment, and getting ready to deliver the training event. It may not be possible to rehearse the whole training course, but taking the time to practice indispensable portions such as manipulating a model or applying effective presentation and facilitation skills will prove beneficial.

9.3 EFFECTIVE PRESENTATION—YOU SHOULD BE NERVOUS!

Why should you—or any trainer or presenter—be nervous? If you care about what you do, you would be nervous about it, because you are naturally concerned whether you will deliver the best outcome. That is regardless of how well prepared or knowledgeable you are on the topic, or how many times you have given the same training or presentation. This kind of nervousness is different from the fear that Derek experienced. The key is to turn that nervousness into positive energy and enthusiasm.

You should care about your purpose—to help your trainees learn—not about trivial matters. So, be yourself and, although your appearance should be professional, do not worry too much if your hair style does not look perfect, your shoes are not polished to a mirror shine, or your pants do not have sharp creases like in the military. Realize that your trainees are there to learn, not to critique you under a microscope in the same fashion as they may scrutinize a movie star or celebrity.

Professional speakers explain in many different ways how to give a successful presentation. Along with prior proper preparation, they boil down to five critical factors summarized in the acronym SPEAK:

Style
Purpose
Emotions
Audience
Knowledge

Presentation and Facilitation 79

9.3.1 Style

Style has several aspects: organization, vocabulary, vocal variety, body language, and visuals.

9.3.1.1 Organization

A well-organized training presentation is similar to a well-written essay. It has an opening, a body, and a conclusion. The so-called universal speech outline—"Tell them what you are going to tell them, tell them, and then tell them what you told them"—is very appropriate for training. Embellish each part with attention-grabbers.

For example, in the opening, several techniques may be used to capture the attention of your trainees:

- Ask a startling or rhetorical question.
- Tell a poignant story or use a relevant quotation.
- Show an image or short video of a dramatic scene or object.
- Explain enthusiastically the benefits this training brings to each trainee.
- Acknowledge a member of the group for significant contribution to the organization.

The body of the presentation is the main theme of the training. Depending on the subject matter and your instructional strategies, consider using some of these patterns of organization at various times during the training:

- *Chronological*: Make use of a time line that illustrates significant events. For instance, if the training is for new employees, you can provide a historic perspective of the organization or the department.
- *Comparative*: This pattern can be used to show the difference between organizations, competitors, products, services, or abstract concepts.
- *Cause-and-effect*: Describe the outcomes of certain events. As an example, in training real estate agents, a trainer may discuss the decline in property values due to high rates of mortgage foreclosures.
- *Topical*: This outline organizes the body of the presentation by topic areas. In a training course of a suite of office productivity software, for example, the topic areas may be word processing, spreadsheet, database, and so on. If there are differences in the degrees of difficulty of learning the topics, arrange them from the simple to the complex.
- *Problem–solution*: The strategy can be to outline a problem and reveal a solution, or ask the trainees to furnish a solution. For example, in training new managers in the retail business, ask them what the solutions are in combating shoplifting.

The conclusion summarizes what has been taught. It is the final opportunity you have to persuade, convince, and motivate your trainees to remember what they have learned, change their behavior, or improve their performance. You want to conclude with a call to action or state the next steps. Just as you grab your trainees' attention at the opening, do the same at the end, using an inspirational story, quotation, or dramatic effect that will be memorable.

9.3.1.2 Vocabulary

"A picture is worth a thousand words" and a rich vocabulary enhances word pictures. Keep in mind, however, that the most beautiful words are meaningless if not understood by the listener. Take into account your trainees' educational and cultural background and length of experience on the job. Avoid jargon that may be incomprehensible or confusing to them. Except for technical terms that are within the scope of the training, use short words. Action verbs and the active voice add impact to your message. Concise sentences facilitate understanding and are less likely to contain grammatical error. Unlike written communication that can be reread over and over again for clarification, the meaning of spoken words should be immediately clear to the trainees. Even if the training is recorded, most trainees would not go back to review the recording.

Be specific in the choice of words. This is important because your trainees should learn exactly what they are supposed to know or do. For example, if you are conducting ergonomics training and want to mention that many design challenges can be overcome using tools and devices that do not cost an arm and a leg, say, "Ergonomics solutions can be inexpensive," not "Ergonomics solutions can be cheap." When you explain abstract concepts or complicated structures, repeat them in different words or phrases to make them clearer and more interesting.

9.3.1.3 Vocal Variety

A trainer who speaks in a monotone can make a fascinating topic boring. A forceful and expressive voice commands attention. Adjust your voice volume, pitch, and rate to add meaning to the content you are delivering (Toastmasters International, 2009, 32).

Diaphragmatic breathing will help you project your voice without causing vocal tension. That means breathing deeply from the abdomen instead of the chest. The quality of your voice will improve too. Vary your voice volume for emphasis. When you lower your voice volume, make sure that trainees in the back of the room can still hear.

Similarly, changing your voice pitch can have dramatic effects. For example, use a high pitch to convey enthusiasm and a low pitch to show thoughtfulness. Individuals' vocal range varies. Do not strain to reach the upper or lower extremes of your range. Your voice should sound natural, sincere, and friendly.

Your rate of speaking should be not too fast for the trainees to absorb what you say and not so slow that it puts them to sleep. Speaking at between 125 and 160 words per minute is best (Toastmasters International, 2009, 33). Compared with variations in volume and pitch, speaking rate is more dependent on the audience. Slow down when the trainees are novices in the field or some of them may not be fluent in the language, also when explaining a complex issue. Clever use of pauses can be effective with any audience. A pause before or after an important statement emphasizes its significance. A pause after posing a question gives the trainees time to ponder the question. When you do not hear an immediate response, you must overcome the urge to provide the answer right away. Give them a few seconds to think. Audible pauses such as "ah," "um," and "you know" should be eliminated.

9.3.1.4 Body Language

Nonverbal communication is as important as verbal communication. Using a variety of body language adds meaning and interest to your message. Body language includes eye contact, facial expression, posture, gestures, and movements. They should match and not contradict your message. Have you played the game where you are required to shake your head when replying "yes" and nod when saying "no"? It feels awkward, at least in most cultures. When used correctly, gestures make your training presentation more impressive and unforgettable. When used incorrectly, they are distractions. The challenge is that what is correct and what is incorrect can be culturally dependent. Chapter 16 discusses circumstances that may arise when training a multicultural workforce. The descriptions in this chapter apply generally to Western culture.

Eye contact means looking a person in the eye. It is not glancing around the room. With a group, look at one trainee for a few seconds and then another trainee in a different part of the room. Pick the persons randomly and be sure to include those in the back. Direct eye contact not only shows confidence but also satisfies your trainees that you are paying attention to them. It helps to maintain their attention and build rapport. It also enables you to find out if they are attentive or comprehending. When you see a puzzled look, you may need to repeat in a different way what you have said.

Facial expressions, shown by movements of your eyes, eyebrows, and mouth, display emotions. In addition to portions of your training when you play a role, your facial expressions throughout the training may subtly affect how your trainees learn. That is because they will emulate your feelings, albeit subconsciously. A pleasant smile is encouraging, whereas a frown that you may not be aware of yourself has the opposite effect.

Your posture affects your credibility. When you are poised, you project confidence and professionalism. Avoid postures that have been given stereotyped interpretations. For instance, crossing the arms at the elbows in front of the body is viewed negatively as indicating that the person is reserved, unfriendly, or self-conscious. It does not matter if that person is in fact just feeling too cold in a room! When you are standing still, a neutral posture with hands on the sides or in front of you, ready to gesture appropriately, is preferred. A good posture also helps to improve voice quality as it allows air to flow freely through the vocal chords.

Most gestures are made with the hands and arms, but they can be made with other parts of the body, such as shoulders. As an example, shrugging the shoulders communicates indifference or giving up (as when one does not know the answer). Hand gestures can show size, shape, and weight if you do not have the actual object. When using hand gestures, be careful not to point your index finger to one trainee or the group. This is considered rude, as suggested in the expression "pointing fingers." Pointing your finger to the ceiling is acceptable as it denotes "Pay attention to what I'm going to say." Hence, a small change in your gesture can result in a substantial difference in meaning. A common mistake of inexperienced presenters is not using gestures or using them in a repetitious manner, which becomes distracting.

Gestures should be purposeful, varied, and natural. If you have a large group, though, you may want to exaggerate your gestures somewhat so that everyone can see.

Body movements that relate to the message add impact but constantly pacing back and forth causes distraction. Other inappropriate movements include repetitively shifting the body weight from one foot to another or fiddling with hair, jewelry, clothes, and suchlike. When you describe an action, you can stand still or you can act it out using body movements. Which is more effective? Using body movements, of course. Moving around during a training presentation serves other purposes as well. You maintain attention of the audience as your trainees' eyes follow you. Moving from one spot to another is complementary to transitioning to a new point or topic. Stepping forward toward the audience helps to emphasize an important point. Avoid turning your back toward the audience as you are moving.

9.3.1.5 Visuals

The selection of visuals for training is discussed in Chapter 6. Visuals can be your friend or your enemy. When handled smoothly, they add impact to your message. When not used properly, they become a hindrance.

How do you employ visuals effectively in your training presentation? The answer depends on the nature, physical attribute, and purpose of the visuals. In all cases, maintain eye contact with your audience and do not speak to an object while you are holding it or standing next to it. If you hold an item in your hand or if it is a large item that stands on the floor, display it to the side of your body so your trainees can see it. If a large item does not need to be referred to repeatedly throughout the training, place it in the corner of the room and set aside a time for the trainees to view it. Visuals should be clearly visible, or at least each trainee should have an opportunity to take a close look. A small article can be passed around, but this can create distraction. If you want to show an old photograph developed on paper, it is better to digitize it by scanning or use equipment such as a document camera to project the picture onto a screen.

Never stand in the light path of a projector—your trainees can see neither your face nor what is on the screen. If you need to point to something on a slide, use a laser pointer. For that matter, do not block the view of any visuals. For example, after you finish writing on an easel pad, stand to its left side, from the audience's view. Why left side? Because you want eye contact with the trainees as you resume speaking, and people generally look from left to right. You may have to get used to doing this because if you are right-handed, you probably have a tendency to stand on the right side so you can use your dominant hand to point to what is written on the easel pad. For the same reason, if you project materials on a single screen, it is preferable for you to stand to the left side of the audience. When two projection screens are used, or there is a screen and an easel pad, you would be in the middle. Also, do not speak while you write on an easel pad. Recruiting a scribe to write is better.

9.3.2 PURPOSE

In public speaking, the purpose of a presentation usually falls into one or more of four categories: to inspire, inform, persuade, or entertain. In training, the first

Presentation and Facilitation 83

three would be your primary purposes. While a training presentation that entertains would improve learning retention and is highly desirable, it is a means, not an end, unless you are training speakers or performers. Maintain focus on the learning objectives. Do not allow the entertaining aspect to eclipse the true purpose of your training.

9.3.3 EMOTIONS

Emotions are contagious. When you are passionate about the topic, your trainees will feel and share your enthusiasm. When you unwittingly display the slightest sign of frustration, disgust, or anger, they will feel it too and their desire to learn will diminish. If you happen to have a bad day, control your emotions and concentrate on the task at hand. Putting your worries aside may actually help you feel better. You should also project confidence without appearing arrogant or condescending.

9.3.4 AUDIENCE

In the process of developing your training plan, you have already analyzed your audience. As you prepare for the training event, consider the size of the group. A large group requires more time for interaction and you need to plan your agenda accordingly. You may also have to recruit a training assistant for safety reasons and to ensure that every trainee will receive attention and assistance as needed. During the training presentation, connect with your trainees through your style and emotions. Watch their reactions and adjust your plan if necessary.

9.3.5 KNOWLEDGE

As the trainer, you would be a subject-matter expert. You may feel an urge to teach the trainees everything you know. Do not succumb to that temptation! You will overwhelm your trainees. Stay on track and adhere to what is relevant to the learning objectives, the training plan, and the agenda. Before the training event, try to think about questions that the trainees may bring up so you can decide in advance (a) how to explain your answers in ways that the group will understand and (b) which types of questions should be deferred because they are diversions.

9.4 EFFECTIVE FACILITATION—WHEN SHOULD YOU STOP PRESENTING?

Training is not just giving a presentation. If you recall, part of the definition of training states, "Training is a communication process involving interaction between trainer and trainee." A trainer is a facilitator who guides the trainees through their learning process, during which two-way communication is crucial. Effective facilitation demands that a trainer knows when to stop presenting and start asking questions, anticipates and responds to questions, and possesses skills in managing individual and group behaviors during the training event.

9.4.1 Asking Questions

Asking a question is a great way to introduce a topic or check for understanding. For example, a trainer is delivering a course on waste management. Before discussing the benefits of recycling resources, the trainer may ask, "What are the reasons for recycling the world's natural resources?" Similarly, a question can be asked at the end of the discussion to let the trainees summarize the points.

Notice that in the example the question states, "What are the reasons ..." It does not ask, "Are there reasons for recycling?" The former is an open-ended question; the latter is a closed-ended question. An open-ended question leads the respondents to think and reason before giving their opinions. Since a closed-ended question requires only a "yes" or "no" answer, not as much careful thought is required. Nonetheless, it is suitable for testing the respondents' memory on facts, such as "Are the federal hazardous waste regulations promulgated under the Resource Conservation and Recovery Act?"

You can direct a question to one trainee or to the group. Asking a specific individual's view is a good way to draw in a shy person. It remedies a situation where a few trainees dominate the discussions. Be tactful so that the trainee asked would not feel being put on the spot. Provide a hint if it appears that the person is struggling to find an answer. To allow everyone to comment, ask a question of the group, or an overhead question. Be patient after you have posed the question and pause long enough for an answer to be forthcoming. People tend to wait for someone else to reply to an overhead question.

9.4.2 Responding to Questions

When one of the trainees indicates that he or she has a question, you should recognize the questioner, by name if known. Listen to the question carefully to make sure you understand it or ask for clarification. Avoid clichés such as "That's a good question." In a large room when the questioner has not used a microphone, repeat the question so everyone knows what is asked. If the question is complex, you may want to paraphrase it or divide it into separate questions. After you have responded, confirm with the questioner that you have answered his or her query.

Some say that it is rude to answer a question with a question, but there is such a question type called *return question*. For example, in the waste management training mentioned earlier, a trainee asks, "Do manufacturing industries object to the burden of regulations?" In response, the trainer may ask a return question, "What do you think?" Such an approach is not to avoid answering the question. Rather, it is to elicit the questioner's thoughts, after which the trainer can add comments or open the discussion to the entire group by asking a relay question, "How do the rest of you feel about it?" This is a combination of the techniques of asking and responding to questions. After the group's remarks, the trainer can fill in gaps that the group may have missed.

It is not necessary to answer all questions right away. Still using the training course on waste management for the purpose of illustration, suppose a trainee has a question about green building design that is not related to the learning objectives.

Although both waste management and green building design are within the broader scope of sustainability, green building design is off-topic. Even if the trainer is very knowledgeable about it, he or she should discuss it with the questioner and whoever else may be interested after the training. The trainer's priority is to focus on the learning objectives.

Another situation for deferring a question is when time is running out. If you are trying to adhere to the schedule and your trainees have more questions, inform them that you can discuss with them after the training.

What if you do not have an answer to a question? Be honest and tell the trainees that you do not know. You are a human being and not expected to know everything on top of your head. Inform them that you will find the answer for them by a certain date and, if the trainees are not a group with whom you have regular contact, verify that you know how to get in touch with them or their representative when you have the answer. Do not forget to get back to them by the date promised.

9.4.3 Managing Behaviors

Behaviors that must be managed range from nonparticipation to overparticipation. For instance, when you ask overhead questions, do you keep getting answers from the same trainees? You are not sure about the ones that offer no comment. Are they shy or are they not paying attention? Do they understand? It is your responsibility, as the trainer and facilitator, to get them involved. As mentioned earlier, you may solicit a particular trainee's opinion. The silent trainees frequently look down at their notes to avoid eye contact with the trainer, hoping that this way the trainer will not call on them. Address an individual by name. Ask open-ended questions and help the person expand on his or her ideas. Your eye contact and facial expression should convey a sense of encouragement. How about those too keen to answer? It is appropriate to request them to hold off and give somebody else a chance to contribute. Be polite but firm.

Most trainees do not exhibit disruptive behavior. Unfortunately, once in a while you may encounter one or two that do. Examples are the know-it-alls and crabs (Toastmasters International, 2008, 41). The know-it-alls may be knowledgeable and have good suggestions to offer. The trouble is that they are unreceptive to others' ideas that are different, possibly including yours. When there is disagreement, avoid arguing with them. Acknowledge that viewpoints may differ and thank them for sharing theirs. The crabs are eternal complainers. They are unhappy about everything, whether it is the room temperature or the chair or the training schedule. They are wet blankets that dampen enthusiasm. Determine if a complaint is valid. If it is, ask them or the group to suggest a solution and carry it out, provided that the solution is feasible. If the complaint is not justified or there is no feasible solution, explain why and move on. Be respectful but do not let the complainers intimidate you.

9.4.4 Providing Feedback

Feedback goes two ways. You receive feedback on trainees' progress in the learning process through their responses to questions, participation in activities, and formative and summative assessments. This feedback enables you to adjust the training plan if needed. You also provide feedback to trainees to help them advance toward mastering the training topic. Such two-way communication is critical for a successful learning experience.

9.5 FUTURE IMPROVEMENT

Effective presentation and facilitation skills are precursors of effective training delivery. These skills can be acquired and improved with practice and experience. How would you know if your skills are improving? Practice in front of a mirror or video record yourself for self-evaluation. Ask a colleague or friend to observe your performance and provide feedback. Some of your trainees may tell you in their course evaluations.

REFERENCES

The Princeton Language Institute, and L. Laskowski. 2001. *10 Days to More Confident Public Speaking*. New York: Warner Books.

Toastmasters International. 2008. *From Speaker to Trainer: Coordinator's Guide*. Rancho Santa Margarita: Toastmasters International.

Toastmasters International. 2009. *Competent Communication*. Rancho Santa Margarita: Toastmasters International.

10 Course Evaluation

10.1 WHY ASK FOR CRITICISM

Course evaluation is used to assess course contents and effectiveness of the trainer. It is part of but different from program validation, which is covered in the next chapter. Course evaluation pertains to experience from a particular training course, whereas program validation pertains to the whole training program that may include aspects such as resource allocation.

Why would you want to ask for criticism? Course evaluation is the critical feedback that enables continuous improvement in content and delivery. Think of it as constructive suggestion, not criticism.

From the trainees' perspective, knowing that they will be evaluating the course upon completion provides an added incentive for trainees to be attentive during the training. Experience shows that many professionals are eager to share their opinions to help improve a training course for future participants. They are also aware of the fact that improvements in general may affect another course they may take later.

From the trainer's perspective, course evaluation motivates the trainer to strive for excellence. After a training course is developed, it is often presented to different groups of trainees at different times. Although every time the trainer must research the specific audience and situation, much of the original materials can probably be reused. A trainer's delivery style is applicable to different courses. Feedback from those who have taken a course can help the trainer identify materials or delivery methods that can be enhanced. Constructive suggestions in course evaluations after each training event should be implemented immediately for the next event whenever possible.

10.2 WHO THE EVALUATORS ARE

Formal course evaluations are typically completed by trainees. Trainee reactions are considered good measures of course characteristics as they relate to trainees' perception of the training environment, particularly trainer style and interactions during the training event (Sitzmann et al., 2008). Reaction measures are especially important when the purpose of the training is to change attitudes since they indicate whether the trainees would be open to change. You should also perform a self-evaluation. Additionally, a third person's viewpoint may be helpful. This person may be a mentor or colleague whom you invite to be an observer. If the person is not familiar with training techniques, he or she can play the role of a trainee and provide feedback as such. Ideally, this person has expertise in training, in which case the comments would be more specific and valuable in helping you refine your training skills.

If you work in the corporate setting, your supervisor may want to observe your training as a component of your performance evaluation. Many continuing education courses are evaluated by auditors to determine if the course meets the criteria for issuing continuing education units. Conference coordinators may evaluate the training courses presented at a conference and compare their evaluations with those of the trainees to get a better idea of the effectiveness of the trainers (Kirkpatrick, 1959a, 7). Regulators may want to evaluate training courses that concern specific technical areas; such evaluations often comprise a review of the training documentation only rather than observation of training delivery. Documentation is discussed in the next chapter.

10.3 WHAT SHOULD BE EVALUATED

Items targeted for evaluation are dependent on the evaluator. Whereas trainees evaluate the course based on their learning experience, the trainer's self-evaluation might include the effectiveness or otherwise of "behind-the-scenes" events that are not evident to trainees. These events range from setting up training aids to ordering refreshments. What auditors or supervisors rate depends on the purpose of their evaluation.

10.3.1 Trainer's Self-Evaluation

From observations during the training event or from the test results, you may recognize opportunities for improvement. Due to Murphy's Law, it is not uncommon for something to go wrong at a training event. When you perform a self-evaluation, blaming yourself for what went wrong is human nature but that could be detrimental. Focus on what you can change and do not dwell on what you had no control over. For example, if an unexpected power outage affected the corporate campus where training took place, you could not have done anything about it. On the other hand, if you needed an electrical extension cord but it was not available, you would ensure that next time you make prior arrangement.

Concentrate on evaluating whether your training plan realized what it intended along the following lines. As shown in Appendix C, space can be included in the training plan for these items of self-evaluation.

- *Learning objectives:* Are the objectives accomplished, as demonstrated by the audience reaction, trainees' evaluations, and test results?
- *Instructional strategies:* How well did your strategies work out for this audience? Will they need modification for a different audience?
- *Training materials:* Were the materials appropriate and used effectively to convey your message?
- *Testing methods:* Are the testing methods valid and reliable? Do the test results appear reasonable?
- *Delivery style:* Did your delivery style engage the audience? Would a different delivery style have worked better for this or another audience?
- *Overall evaluation:* Have *you* learned any lesson from this training—what can be done to make future training better?

Course Evaluation

10.3.2 SUPERVISOR'S OR AUDITOR'S EVALUATION

Generally a supervisor or an auditor will have an evaluation form from his or her organization for this purpose. If your training will be evaluated by your supervisor or an auditor, it would be extremely helpful for you to obtain a copy of that form or find out the evaluation criteria before designing the course to ensure that the criteria will be met or exceeded.

Areas that are commonly evaluated by supervisors or auditors include the following:

- *Course materials*: Were the materials relevant to the course description and learning objectives? Were they organized in a logical flow? Were they prepared with care that reflects professionalism?
- *Instructional strategies*: Did the trainer effectively use a variety of strategies at different stages of course delivery? Were the strategies appropriate for the learning objectives and the particular audience?
- *Training aids*: What training aids were used? Did they seem to have been planned with careful forethought to support the learning objectives? Did the trainer handle the training aids smoothly?
- *Delivery style*: Did the trainer use language that was suitable for the audience and promoted understanding? Were the trainer's voice volume and quality appropriate for the size of the audience and contents of the course materials?
- *Audience engagement*: Did the trainer capture and maintain audience interest? Were trainees encouraged to actively participate in discussions and activities? Was the trainer's demeanor approachable? How well did the trainer respond to questions?
- *Time management*: Was an agenda provided and followed, and adjustments made when justified? Did the trainer start and end the training event on time? How did the trainer make the best use of the available time?

10.3.3 TRAINEE'S EVALUATION

This is your opportunity to elicit the trainees' thoughts about their learning experience to help you improve future training. Positive evaluations, albeit subjective, may indicate that trainees have paid attention and learned what they have been taught (Kirkpatrick, 1959b, 21).

When the training is one-on-one, as is often the case in on-the-job training, the evaluation is likely to be informal and oral. In other settings, course evaluation can be oral or written. Oral evaluation may be solicited during the question-and-answer or wrap-up period toward the end of the training course, or informally during breaks. If time permits, interviews may be conducted after the training. Questions asked of the trainees can proceed along the same lines as in a written evaluation. The written format has the advantage of facilitating record keeping and respondent anonymity. It usually takes the form of a survey distributed and collected manually or electronically. Thoughtful design of the survey would help gather useful information.

10.4 HOW TO DESIGN A COURSE EVALUATION SURVEY

Course evaluation surveys used to be handed out at the beginning of an in-person training event, with the request that trainees complete the survey at the end of the course before they leave. Nowadays, more and more evaluations are completed electronically. Whenever the evaluation form is not distributed at the beginning of the training event, it is advisable at that time to show the trainees the questions they would be asked in the survey. This approach helps them in providing thoughtful responses.

10.4.1 QUESTION DESIGN

The advent of websites such as SurveyMonkey® has given rise to a misconception that anyone can write survey questions. Good survey design takes advantage of specialized knowledge and skills in research methods. While more on this topic will be discussed in Chapter 14, a few useful guidelines are to (a) include both closed-ended questions and open-ended questions with spaces for write-in comments, (b) avoid question content or sequence that may skew the answers, and (c) provide a space for the evaluator to include his or her name, if desired, so that the individual may be contacted for clarification of his or her responses.

10.4.1.1 Closed-Ended and Open-Ended Questions

A closed-ended question leads to a yes/no answer or uses a rating scale. It restricts the answer to a narrow range, as in the following examples:

- Did the trainer respond to questions effectively?
- On a scale of 1 to 10, 1 being least satisfied and 10 being most satisfied, please rate how satisfied you are with the manner in which the trainer responded to questions.

An open-ended question allows the evaluator to provide insight that may not be expressed in the closed format. Below are examples of open-ended questions that try to gain a deeper understanding of trainees' opinions without constraining them to predetermined answers.

1. Which topic(s) do you think should be added to this course in the future?
2. Please make one or more suggestions as to how this course can be improved.

Both types of questions have advantages and disadvantages; therefore, it is a good idea to use them in combination.

A closed-ended question typically has five to ten answer choices. The answers are easy to tabulate manually or electronically, saving time and cost. A rating scale, or answer choices that can be converted into a rating scale, simplifies statistical analysis. The opinions given may be influenced by the construction of the questions and possible answers. Debate goes on whether there should be an odd number or even number of answer choices when the question measures the range of one variable, such as the degree of satisfaction. The following question is an example.

Question:
> How satisfied are you with the boxed lunch included in this training event?
> Odd number of answer choices:
> "Very Satisfied"—"Satisfied"—"Neutral"—"Dissatisfied"—"Very Dissatisfied"
> Even number of answer choices:
> "Very Satisfied"—"Satisfied"—"Dissatisfied"—"Very Dissatisfied"

On the one hand, an odd number of choices that includes a "neutral" answer enables evaluators to express a neutral opinion if that is truly what they think. On the other hand, it may encourage too many "neutral" answers just because it is easier for an evaluator to check this answer than to spend time thinking which other answer choice is more appropriate. For this reason, some survey designers are in favor of an even number of answer choices, which would eliminate "neutral" answers. Doing this, though, may result in inaccuracies when some evaluators are forced to pick options that do not reflect their opinions. Either way, answer choices given in the range should be exhaustive and should not overlap.

With open-ended questions, evaluators' answers still may be influenced by how questions are framed but since the range of answers is not limited, they are not as likely to be skewed as when closed-ended questions are asked. Open-ended questions also show the evaluators that you care about their opinions. Text responses are more difficult to tabulate. They must be read individually and categorized. Their meanings may be unclear and are subject to the interpretation of the reader—in this case the trainer; hence, it is a good practice to request the names of evaluators in case it is necessary to seek clarification of their responses. Text analytics software is available from web-based providers of survey solutions or may be purchased as a stand-alone product or part of a statistical analysis package. Text analytics alleviates the time-consuming process of manual categorization and makes text responses quantifiable (IBM Corporation, 2010, 11). Features range from simple statistics-based keyword search to sophisticated data mining that can reveal patterns and relationships (Grimes, 2010, 2).

Answers to open-ended questions may explain the "why" behind the answers to closed-ended questions (Grimes, 2010, 2). That is an advantage of using both types of questions in one evaluation. A balance between the two kinds of questions is needed since too many open-ended questions may discourage evaluators from responding altogether due to the time and thought they must put in.

10.4.1.2 Leading and Loaded Questions

A leading question is framed so as to guide the respondent in the reply (Merriam-Webster Online, 2013); the question contains the desired answer. Consider the following two questions:

1. Did you attend this training course because your job required that you attend?
2. Why did you attend this training course?

The first question is a leading question as it suggests the answer—the evaluator's job requirement. The second question is not a leading question as many answers are possible. For instance, an evaluator may attend a training course to acquire the latest information in the field or network with other trainees.

A leading question may exist in a closed-ended question format, when a range of answer choices is not evenly distributed throughout the range. In a range where "Excellent"—"Very Good"—"Good"—"Fair"—"Poor" are the five choices, answers are likely to be more positive than negative. A range of answer choices that is evenly distributed would be "Excellent"—"Above Average"—"Average"—"Below Average"—"Poor."

A loaded question contains a presumption and tries to confine responses to suit the questioner's purpose. Again, consider two examples where the second is not a loaded question but the first is because it assumes that the techniques are useful to the evaluator.

1. When will you begin to apply the useful techniques taught in this course in your job?
2. If any techniques taught in this course are helpful to you, when do you expect to use them in your job?

Leading and loaded questions have inherent biases and should be avoided in course evaluations and surveys.

10.4.1.3 Word Usage

Sometimes the effects are more subtle, as when words such as "always" or "never" are used. Avoid these words unless you truly mean them. Double negatives are confusing. Presenting the same question in a positive or negative form may also skew the responses (Pedhazur and Schmelkin, 1991, 137).

10.4.1.4 Question Sequence

The order in which questions are placed within an instrument may affect the responses (Pedhazur and Schmelkin, 1991, 140). Turocy (2002, S-177) recommends grouping questions with similar content together provided that earlier questions would not affect the responses to later questions. Do not embed a question within another question.

10.4.1.5 Trainee/Evaluator Anonymity

Results of research on the impact of anonymity on the response rate of surveys are inconclusive (Faria and Dickinson, 1996, under "Anonymity"). It has been recommended earlier that the name of the evaluator should be optional. This approach is suggested as it has no negative effect. It may encourage trainees to respond, knowing that they are not individually identified, especially when the size of the group is small or you had a lot of interaction with each trainee.

10.4.2 RESPONSE RATE

The ease of conducting electronic surveys has resulted in people being bombarded by requests to complete surveys every week or several times a week, from trainers and conference organizers to store managers. It is understandable why some people would tune out all requests and not want to fill out any survey. Others may do it selectively and may be more inclined to complete a survey that is meaningful to them, such as a course evaluation.

A course evaluation form that is brief and easy to fill out tends to have a higher response rate than one that is time-consuming to complete. Still, it should cover various aspects of the course from design to delivery. At an in-person training event, the response rate may be increased if a monitor collects the form as trainees leave the room after the training or reminds them to place the form into a drop box at the exit. For virtual training, completion of the course evaluation can be made mandatory before issuance of a certificate of completion, or an incentive may be given to trainees that provide evaluation, such as a discount for the next training registration. When the evaluation is completed electronically, you can help the trainees by letting them know, at the beginning of the survey, the estimated time needed to complete the evaluation.

10.4.3 Timing

Most course evaluation data are collected immediately or soon after the training event while memories are fresh. If several training courses comprise a series, using information from evaluations of the earlier training courses can help improve later courses in the series (Kirkpatrick, 1959a, 5). Occasionally, a follow-up survey is performed 3 to 6 months later to determine if there is any change in trainees' perception of the relevancy of the training to their job. If a follow-up study is conducted, the questions asked would be different from the initial survey but the same principles of design would apply. If the workload of the trainees varies in the course of a year, avoid follow-up surveys during peak periods.

10.5 WHICH DATA ARE RELEVANT

A pitfall for some incidental trainers is the tendency to take every rating or comment received in the course evaluation too personally. They become discouraged by low ratings and negative comments. It is not untypical for evaluations of the same course by different evaluators to comprise a spectrum of responses that range from the negative to the positive. You should remember that the evaluations represent individual perceptions and may not be objective, even though subjectivity may be reduced by the way survey questions are framed, as discussed earlier in this chapter and in Chapter 14. Do not let negative comments dampen your spirits. Instead, select and adopt constructive suggestions.

For example, if one trainee out of thirty gives the course a low rating, you may want to consider, "Has this person also provided any supporting evidence or specific suggestion?" If not, the evaluation does not contribute to course improvement. In contrast, if several trainees comment, say, that more time should be devoted to a certain topic covered in the course to examine the subject matter thoroughly, you may want to modify the agenda accordingly the next time around.

A word of caution is in order. Welty (2007, 68) points out "the possibility of documented negative judgments" that applies to training conducted for regulatory compliance. If you include a question in the course evaluation that solicits suggestions on how the course can be improved, you should be prepared to take corrective steps, within reason, based on the feedback you have received; otherwise, a regulator reviewing the training documentation may question why no remedial action has been taken.

10.6 WHAT ELSE MUST BE ASSESSED

You have delivered the training course. Trainees have been tested. The course has been evaluated. What else needs to be assessed? The training program itself must be validated. Except in very small organizations, incidental trainers are not responsible for all employee training in the organization. You may have a training program within your technical area, department, or business unit. For instance, if you are the safety officer, you may be responsible for all safety training that covers a variety of topics in health and safety. Regardless of the breadth of your training program, the program's effectiveness should be validated to demonstrate its value to the organization. Program validation is discussed in the next chapter.

REFERENCES

Faria, A. J., and J. R. Dickinson. 1996. "The effect of reassured anonymity and sponsor on mail survey response rate and speed with a business population." *Journal of Business and Industrial Marketing* 11 (1): 66–76. doi:10.1108/08858629610112300.

Grimes, S. 2010. "Text Analytics in the BI Ecosystem." White Paper. Sybase Inc. http://www.sybase.com/files/White_Papers/Sybase_IQ_Text_Analytics_in_the_BI_Ecosystem_wp.pdf.

IBM Corporation. 2010. "Analyzing Survey Text: A Brief Overview." Working Paper YTW03100-USEN-02. IBM Corporation. http://public.dhe.ibm.com/common/ssi/ecm/en/ytw03100usen/YTW03100USEN.PDF.

Kirkpatrick, D. L. 1959a. "Techniques for evaluating training programs." *Training Directors* 13: 3–9.

Kirkpatrick, D. L. 1959b. "Techniques for evaluating training programs: Part 2—Learning." *Training Directors* 13: 21–26.

Merriam-Webster Online, s.v. "Leading Question," accessed February 18, 2013, http://www.merriam-webster.com/dictionary/leading%20question.

Pedhazur, E. J., and L. P. Schmelkin. 1991. *Measurement, Design, and Analysis: An Integrated Approach*. Hillsdale: Lawrence Erlbaum Associates.

Sitzmann, T., W. J. Casper, K. G. Brown, K. Ely, and R. D. Zimmerman. 2008. "A review and meta-analysis of the nomological network of trainee reactions." *Journal of Applied Psychology* 93 (2): 280–295. doi:10.1037/0021-9010.93.2.280.

Turocy, P. 2002. "Survey research in athletic training: The scientific method of development and implementation." *Journal of Athletic Training* 37 (4 Supplement): S-174–S-179.

Welty, G. 2007. "Developing assessments of trainee proficiency." *Journal of GXP Compliance* 12 (1): 64–73.

11 Program Validation and Continuous Quality Improvement

11.1 AN INTEGRATED TRAINING PROGRAM

Training should be an integrated program, not a series of unrelated training events. If an organization only trains its employees every time it has a performance concern or has received a regulatory citation, the organization has no training program.

From an organizational perspective, when designing a training program, the process of training needs analysis should be used to determine the requirements for every job category and a plan should be devised to fulfill those needs, taking into account regulatory mandates, performance requirements, employee development, and the need for refresher courses.

The same principles apply to a business unit or department within an organization. In a technical support department, for example, team members that are new to the organization would participate in new employee orientation provided for all new employees, as well as specific orientation in the department. Entry-level technical support personnel assigned to handle initial customer requests would be trained on the systems or equipment that the department supports, including the most common problems that are the reasons for customer requests—even if a new team member had done similar work in another organization, protocols might be different. As the team members grow in experience and proficiency, they would be trained on advanced features, problems, and solutions to support complex issues that customers might encounter. Refresher courses at each level would be appropriate from time to time as reminders and to teach "tips and tricks" that could improve efficiency. Besides technical training, team members might attend courses in customer service, communication, and interpersonal skills to round off training needs for all facets of their job. Additionally work force safety training would take place to comply with regulatory standards. These different courses would comprise the department's training program and the training programs of all departments would make up the organization's overall training program. Some of the courses that have shared learning objectives such as general new employee orientation may be combined for all departments and conducted by the human resources department, while courses specific to a department are usually delivered by staff of the department, often incidental trainers. The services of third-party consultants may also be engaged. Regardless of who the trainers are, the training program for a department and organization should be validated.

11.2 PROGRAM VALIDATION PURPOSES

The twofold purpose of program validation is to demonstrate the effectiveness of a training program in enhancing performance and to explore opportunities for improving the program.

Establishing a training program's organizational impact is a must to garner support for the program. The value of the program can be tangible or intangible, direct or indirect, and immediate or delayed. In the technical support department mentioned earlier, trained employees are more efficient, which results in higher productivity, lower cost per service call, and quantifiable financial gain. Improved customer satisfaction, albeit qualitative, cannot be dismissed. Superior performance of the department helps boost the reputation of both the department and the organization as a whole, so indirect benefits go beyond the department. Outcomes such as cost savings may be realized within a short time, whereas increased business from a favorable corporate image may take longer to materialize.

Evaluations of individual training courses provide data that serve as the basis for refining the courses. Similarly, the process of validating the training program helps the trainer or training department discover areas for improvement of the program. Some of these areas may be the sequencing of training topics, scheduling of the courses, or utilization of resources.

11.3 PROGRAM VALIDATION CRITERIA

In a series of publications, Kirkpatrick explained a four-level model for evaluating training programs.

- *Level 1—Reactions:* Trainees' opinions about the training (Kirkpatrick, 1959a)
- *Level 2—Learning:* The increase in knowledge, improvement in skill, or change in attitude, as demonstrated during the training (Kirkpatrick, 1959b)
- *Level 3—Behavior:* Transfer of what is learned to the job (Kirkpatrick, 1960a)
- *Level 4—Results:* Outcome of the training program in terms of benefits accrued to the organization (Kirkpatrick, 1960b)

Other authors have suggested variations of this model or new models. For example, Phillips (1996b) proposes a five-level model where Level 5, Return on Investment, measures whether the training value exceeds the training cost. In a meta-analysis of the relations among training criteria, Alliger et al. (1997) used an augmented framework that distinguishes affective reactions from utility judgments in the measure of reactions, and separated learning into immediate knowledge, knowledge retention, and behavior or skill demonstration. The basic principles, however, remain the same and the Kirkpatrick model is still widely accepted and used by trainers and educators.

Moving from level 1 to level 4 evaluations requires more and more resources to collect data. Many organizations do not go beyond level 1. Some organizations evaluate a percentage of their training courses, with lower percentages at the higher levels. For example, they may conduct level 2 evaluations for 50% of the courses,

Program Validation and Continuous Quality Improvement

level 3 evaluations for 20% of the courses, and level 4 evaluations for 5% of the courses. It is better to use such a system than no evaluation at levels 2 to 4 at all.

The first three levels of evaluation—reactions, learning, and behavior—are the same criteria that are measured in course evaluations. Those data can be aggregated and extrapolated from the microscale of an individual training course to the macroscale of a training program. Additional data collection and analyses may substantiate the findings further.

11.3.1 LEVEL 1—REACTIONS

In addition to the course evaluation, a separate survey of trainees' and their supervisors' reactions can be conducted to obtain feedback on the training program as a whole. Such a survey may seek answers to the following and other questions:

- Does the training program contribute to improvement in performance? What other factors might have caused any change in performance?
- Are the courses delivered or accessible to the trainees at convenient times and places to minimize work interruptions?
- Are courses scheduled in a way that trainees can attend in a sequence that matches progressive changes in their job responsibilities?
- What other courses should be made available to enhance trainees' job skills?

This type of data would not be collected after each training event. A semiannual or annual survey is sufficient.

11.3.2 LEVEL 2—LEARNING

The original Kirkpatrick model associates the learning criterion in level 2 to knowledge, skill, and attitude demonstrated during the training. At first sight, it seems that such information would be evident from the testing and assessment pursuant to the course, and test results of all training events can be analyzed to detect trends:

- If pretests and posttests are used, how do the differentials between pretest and posttest scores compare among different training courses?
- Do different training events of the same course have similar results?
- Is there a unit or job category that consistently has high or low test scores?

In practice, obtaining reliable results to prove that the gain in test results is due to the training is not easy without a controlled experiment, which is discussed in Section 11.4.4.

11.3.3 LEVEL 3—BEHAVIOR

The purpose of training is to transfer knowledge, teach skills, or transform behavior in support of an organization's goals. Training is unsuccessful if what is taught is not applied to the job, no matter how satisfied the trainees are about the training program or how high all test scores are. Transfer of learning from training to the job should be appraised systematically and statistically on a before-and-after basis at least 3 months after the training (Kirkpatrick, 1960a, 14). It can be studied through surveys, observations, and experiments in which various stakeholders participate. Each study can be focused on trainees of one course or trainees of multiple courses in the training program. The question to be answered is whether a target performance level has been reached (Sackett and Mullen, 1993, 616). The target performance level would have been decided in the training needs analysis.

Data may be available also from routine employee performance appraisals. However, it is a common practice for organizations to implement a 360-degree evaluation in employee performance appraisals, and Shore et al. (1998, 291–292) find that the rating of an employee's self-assessment may be a source of potential bias in the supervisor's rating of the employee; therefore, this type of performance appraisal should be used with caution. Instead, a performance contract may be expedient, when measurable training goals are established prior to training and goal achievement is verified after training. Action plans are even more powerful, where trainees document in a task or project their application of what they have learned (Phillips, 1996a, 47).

11.3.4 Level 4—Results

While benefit to the organization is the ultimate goal of training, many training programs are not evaluated at level 4 or results are not quantified, probably because level 4 is the hardest to measure. Notwithstanding the time and effort required, level 4 is an integral part of program validation and cannot be neglected, especially when the training program must compete with other organizational priorities for limited resources that include not only funding but also the staff's time and attention.

Follow-up studies that are designed carefully and that encourage trainees and other stakeholders to be candid can demonstrate whether the training program has produced practical benefits. After attending a series of training courses and applying the knowledge and skills they were taught, trainees may be in a better position than immediately after each course to offer constructive suggestions to improve the training.

Results may be measured in increased productivity, improved quality, higher profitability, decreased employee turnover, reduced adverse events, and similar parameters. The challenge is how the metrics can be measured so that any change correctly reflects the result of a training program and is not contaminated by confounders. Controlled experiments are most desirable but hardest to carry out in the workplace. Longitudinal studies can measure results in the longer term but are also difficult to design; cause and effect are hard to infer due to the likelihood of more confounders in a long period. Since immediate results have the most influence on organizational support, longitudinal studies will not be discussed further.

11.4 PROGRAM VALIDATION TOOLS

Using a combination of various tools gives a more accurate picture of the effectiveness of the training program. In analyzing data, trends are more important than isolated instances.

Program Validation and Continuous Quality Improvement

11.4.1 Course Evaluation and Test Result

Course evaluations from various training events can be consolidated to detect trends common to more than one training course and opportunities for improvement of the training program. Test results provide direct feedback. They may also reveal issues that are not discovered in reaction surveys. For example, by tracking and comparing test results on an ongoing basis, it is found that one group of trainees regularly scores lower than other groups regardless of the trainer or the course. What could be the reason? It is possible that this group consists of new hires that have insufficient prerequisite skills, in which case a new training needs analysis may determine if remedial training would alleviate the situation. It is also possible that the group is in a business unit that has so much pressure of work that its employees' minds are preoccupied with other priorities, instead of focusing on learning, when they are attending training courses. The trainees and their supervisors may not disclose readily such organizational issues in a survey and you may have to "play detective."

11.4.2 Reaction Survey

Some authors suggest that qualitative trainee reactions do not relate to learning or performance (Alliger et al., 1997, 353). Others consider trainees' posttraining self-efficacy a good predictor of knowledge transfer (Sitzmann et al., 2008, 289). According to Kirkpatrick (1959a, 8), trainees only receive maximum benefit from the training if they like it. Gordon (2007) reports that several well-known consulting firms favor trainee or stakeholder reaction measures. These firms make training decisions based on trainees' responses to questions such as whether the training is a good use of the trainees' time and whether trainees would recommend the training to their peers. In addition, experiments are often difficult to set up in the workplace, so reaction surveys are used to investigate possible confounders. For instance, questions can be included to directly ask if trainees had other experiences concurrently with the training program that might affect training results (Sackett and Mullen, 1993, 621). Since reaction surveys are relatively inexpensive, they are among the most popular tools of program validation.

Chapter 10 discusses question design in course evaluations. The same concerns apply to survey questions for program validation. Leading and loaded questions introduce bias and should be avoided. Response rates for periodic surveys that are separate from course evaluations may be lower than response rates for course evaluations, as some trainees may consider these surveys less important or have simply overlooked the requests. One way to alleviate this problem, if training events take place at regular intervals throughout the year, is to distribute semiannual or annual program surveys alongside course evaluations distributed at training events that occur at about the same time of the year. Chapter 14 discusses additional techniques aimed at producing meaningful surveys.

Who should participate in surveys besides the trainees? The trainees' supervisors and peers are good sources of information. Opinions of trainees' subordinates may be valuable (Phillips, 1996b, 32). If the training program is directly related to customer service, customers may be invited to participate.

A special form of interview is the focus group, a qualitative research method that is frequently confused with discussion groups or reaction panels. A focus group is a structured group interview (Morgan, 1998, 1). It can be organized for any group of stakeholders but is most often used to gain customer feedback. A trained moderator guides participants in discussions on questions raised on predetermined topics. Qualitative data are derived from the discussions and analyzed. When developing the questions, the moderator may select one of two strategies: the topic guide or the questioning route (Krueger, 1998b, 9–12). The topic guide consists of a list of words or phrases that serve as reminders of the topics to be discussed. The questioning route comprises a list of questions written in complete sentences. For example, over the last 6 months the customer service team has attended a training program that included courses on technical skills in computer proficiency and soft skills in dealing with customers. A focus group is convened to determine if customers perceived any change in the quality of service. The same item to be discussed might be listed in different ways depending on the moderator's strategy:

- Topic guide: Change in service quality.
- Questioning route: Think about the last two times when you contacted the customer service department. How were your experiences similar or different on the two occasions?

The questioning route generally works better for individuals who are not professional moderators because it ensures that the question is asked in the way it is planned. Many analytical techniques used in other types of qualitative research can be applied to analyze focus group data. An important distinction to keep in mind is that focus group data are influenced by group dynamics that are absent from individual interviews (Krueger, 1998a, 20).

11.4.3 Observation

Observations on the job can provide insight as to whether trainees are applying what they have learned in the training program to their daily work. Examples are observations by supervisors or trainers. Standardized performance checklists should be used for observations to minimize the subjectivity of the qualitative data. The observation should be done discreetly; otherwise, the Hawthorne effect may come into play whereby employees might modify behavior because they are being observed. Objective data can be collected through measurement of a suitable performance metric. For instance, in an inside-sales call center, by how much has the average time to handle a customer order changed after agents have been trained on the advanced features of the order system?

11.4.4 Controlled Experiment and Quasi-Experiment

In a controlled experiment, a study group and a control group have the same job activities and other characteristics. The study group completes the training program and the control group does not. Group membership is randomly assigned. In practice, random assignment is hard to achieve and a quasi-experimental design might be used whereby trainees are assigned based on, say, team membership on the job. In a department that has two teams doing similar work, only one team goes through the training program and is the experimental group. The other team serves as the control group. Without random assignment, it is important to compare the experimental and control groups to verify that they have similar attributes. Care must also be taken to ensure that the groups are not exposed to other factors that may influence the outcome. Such factors may include supervisor feedback and visual display of performance results that are motivational (Komaki et al., 1980, 268). Before the training intervention, baseline measurements of one or more outcome variables, such as productivity, are taken for each group. After the training intervention, the same outcome variables are measured again. Statistical analysis determines if there is any significant difference within and between the groups.

Experiments are better suited for training programs that aim at enhancing performance. For training programs that must meet regulatory requirements, it would be difficult for the employer to justify why training has been given to one group of employees only, unless the control group is also trained within a reasonable time after the experiment.

Carrying out these experiments in the workplace calls for careful planning. Sample size must be adequate to achieve statistical power. Results should be interpreted with caution. For example, the fact that a training program has produced a statistically significant improvement in an outcome variable in the experimental group does not prove whether a required level of performance has been reached. Nor is statistical significance equivalent to practical significance. For instance, an improvement of one-half standard deviation may not be consequential for practical purposes. Despite these drawbacks, if an experiment is designed and conducted properly, the results have high validity and reliability. They can provide evidence that improved performance is most likely due to the training and not other causes.

11.5 DOCUMENTATION

It is appropriate at this point to mention that program validation studies, as well as other details about the training courses, should be properly documented. Documentation should include the following for each training course and program, as applicable:

- Training needs and task analyses
- Training plan, including justification for the criteria for successful course completion
- Training event agenda
- Course materials
- Record of trainee attendance
- Test and assessment and the results
- Course evaluations
- Program validation study design, data collection, and analyses

"What's not written doesn't exist" cannot be emphasized enough. Training documentation is necessary for certain training courses that are required by law, such as those pertaining to the safety and health of employees, and there are other reasons why documentation is highly desirable. Documentation substantiates adherence to internal policies and expectations regarding employee training. Documentation helps preserve consistency of a training course when delivered multiple times or by multiple trainers. This is especially helpful when, as an incidental trainer, you may not be the only person delivering a particular training course or you may not do it often enough to remember all the details every time. Documentation facilitates comparison between training program outcomes and organizational goals so that discrepancies can be identified and rectified through a process of continuous quality improvement (CQI).

11.6 CONTINUOUS QUALITY IMPROVEMENT (CQI)

CQI is vital to the success of any program or organization. CQI concentrates on the dimensions of quality—input, process, and outcome—and attempts to achieve better outcomes by improving inputs and processes. In the context of a training program, the inputs are resources that include the time and effort of trainers, trainees, and support personnel; supplies; equipment; hardware; software; and facilities. Several processes are involved. They are the processes of designing, developing, and delivering the courses, as well as the learning process itself. The outcome is the result of the training, whether it is measured in test results, job performance, or other metrics. Inputs and processes in the training program should be continuously monitored and evaluated, with the objective of seeking opportunities to improve outcomes. The outcomes must also be monitored to ensure that they meet or exceed organizational goals and stakeholder expectations. In this regard, stakeholders include trainees, trainers, internal and external customers, and all levels of management that are concerned with the particular operational area for which the training program is designed.

Draugalis and Slack (1999) provide a good example of applying the techniques of CQI and the well-known Shewhart/Deming PDCA (plan–do–check–act) cycle in improving instructional strategies. They use the FOCUS-PDCA approach in pharmaceutical education. This approach, developed in 1988 and 1989 by Hospital Corporation of America (quoted in McLaughlin, 2012, 5), consists of the following steps in an iterative process:

- **F**ind a (learning) process to improve.
- **O**rganize a team that knows the process.
- **C**larify current knowledge of the process.
- **U**nderstand causes of process variation.
- **S**elect the process improvement.
- **P**lan improvement.
- **D**o improvement.
- **C**heck data for effect of the change.
- **A**ct to hold the gain and continue improvement.

Suppose you received outstanding course evaluations after a training event. Soon afterwards you observed the trainees on the job and they were doing it in the right way as you taught them. You followed up again in six months' time. To your dismay, they have reverted to doing things in their old, wrong way. Relapse reflects on training program effectiveness. As you learned during needs assessment, performance issues may or may not be due to lack of knowledge or skill. You will need to pinpoint the root cause to determine if the relapse can be prevented by changing the learning process. Several barriers that hinder transfer of learning to job performance have been identified: lack of fundamental knowledge, confidence, time, motivation, supervisor encouragement, and peer support (Marx, 1982, 441; Noe, 1986, 746; Thomas, 2007, 5–6). Some of these barriers are beyond your control. By addressing those that you can control, you can improve the training program and future results. If the barrier is the lack of fundamental knowledge, for instance, you may want to reconsider the prerequisites applicable to the training courses. Another strategy that might be suitable is to motivate trainees by involving them early in the design process of the training (Thomas, 2007, 6). Brinkerhoff and Apking (2001, 1) suggest that the key to achieving training effectiveness is to incorporate performance improvement strategies into the learning process.

11.7 ORGANIZATIONAL IMPACT

It is not enough to conduct and document program validation studies and implement CQI. A training program is sustainable only if stakeholders are convinced that its goals are aligned with the organization's goals that include budgetary and investment objectives. To obtain and maintain organizational support, trainers must be able to present the value of training in a business case using financial and other analytical tools. These are discussed in the next chapter.

REFERENCES

Alliger, G. M., S. I. Tannenbaum, W. Bennett, Jr., H. Traver, and A. Shotland. 1997. "A meta-analysis of the relations among training criteria." *Personnel Psychology* 50 (2): 341–358.

Brinkerhoff, R. O., and A. M. Apking. 2001. *High Impact Learning: Strategies for Leveraging Business Results from Training.* Cambridge: Perseus Publishing.

Draugalis, J. R., and M. K. Slack. 1999. "A continuous quality improvement model for developing innovative instructional strategies." *American Journal of Pharmaceutical Education* 63 (3): 354–358.

Gordon, J. 2007. "Eye on ROI?" *Training,* May, 43–45.

Kirkpatrick, D. L. 1959a. "Techniques for evaluating training programs." *Training Directors* 13: 3–9.

Kirkpatrick, D. L. 1959b. "Techniques for evaluating training programs: Part 2—Learning." *Training Directors* 13: 21–26.

Kirkpatrick, D. L. 1960a. "Techniques for evaluating training programs: Part 3—Behavior." *Training Directors* 14: 13–18.

Kirkpatrick, D. L. 1960b. "Techniques for evaluating training programs: Part 4—Results." *Training Directors* 14: 28–32.

Komaki, J., A. T. Heinzmann, and L. Lawson. 1980. "Effect of training and feedback: Component analysis of a behavioral safety program." *Journal of Applied Psychology* 65 (3): 261–270.

Krueger, R. A. 1998a. *Analyzing and Reporting Focus Group Results.* Vol. 6 of *Focus Group Kit.* Thousand Oaks: SAGE Publications.

Krueger, R. A. 1998b. *Developing Questions for Focus Groups.* Vol. 3 of *Focus Group Kit.* Thousand Oaks: SAGE Publications.

Marx, R. D. 1982. "Relapse prevention for managerial training: A model for maintenance of behavior change." *Academy of Management Review* 7 (3): 433–441. doi:10.5465/AMR.1982.4285359.

McLaughlin, C. P. 2012. "Continuous quality improvement using plan, do, study/check, act (PDSA/PDCA) and quality-improvement tools." In *Implementing Continuous Quality Improvement in Health Care: A Global Casebook,* edited by C. P. McLaughlin, J. K. Johnson, and W. A. Sollecito. Sudbury: Jones & Bartlett Learning, pp. 1–30.

Morgan, D. L. 1998. *The Focus Group Guidebook.* Vol. 1 of *Focus Group Kit.* Thousand Oaks: SAGE Publications.

Noe, R. A. 1986. "Trainees' attributes and attitudes: Neglected influences on training effectiveness." *Academy of Management Review* 11 (4): 736–749.

Phillips, J. J. 1996a. "ROI: The search for best practices." *Training and Development* 50 (2): 42–47.

Phillips, J. J. 1996b. "Was it the training?" *Training and Development* 50 (3): 28–32.

Sackett, P. R., and E. J. Mullen. 1993. "Beyond formal experimental design: Towards an expanded view of the training evaluation process." *Personnel Psychology* 46 (3): 613–627.

Shore, T. H., J. S. Adams, and A. Tashchian. 1998. "Effects of self-appraisal information, appraisal purpose, and feedback target on performance appraisal ratings." *Journal of Business and Psychology* 12 (3): 283–298.

Sitzmann, T., W. J. Casper, K. G. Brown, K. Ely, and R. D. Zimmerman. 2008. "A review and meta-analysis of the nomological network of trainee reactions." *Journal of Applied Psychology* 93 (2): 280–295. doi:10.1037/0021-9010.93.2.280.

Thomas, E. 2007. "Thoughtful planning fosters learning transfer." *Adult Learning* 18 (3/4): 4–8.

Part 2

Training, Like the Pros

12 Gaining Organizational Support

12.1 UNDERSTANDING VALUE AND SUPPORT

You have collected data from various sources and validated the effectiveness of the training program. Would such information not speak for itself and prove that the training program is valuable?

Many programs are valuable but few organizations, whether private, public, for-profit, or nonprofit, have unlimited resources. Competing programs must be prioritized and the typical questions asked in the decision-making process are these:

- What does this program contribute to our organization's critical success factors?
- What are the consequences if we do not have this program?
- What are the alternatives, if any?
- How much are the costs and benefits?
- What is the return on investment (ROI)?

It is generally accepted that some kind of training is needed. The alternatives may be to conduct training in-house or to outsource the function. Either way, the amount of resources allocated to the training program must be justified and costs, benefits, and ROI must be assessed.

Organizational support goes beyond funding. A governing board and executive management that are truly committed to a program give the individual or team in charge of that program high status and visibility within the organizational hierarchy. They also demonstrate support by action, not lip service. In the case of an organization-wide training program, for example, a chief training officer may be appointed and report to the chief executive officer, as opposed to a training coordinator hidden in a cubicle in the human resources department. Visibility is both physical and psychological and also dependent on how the team interacts with other employees. If the training team members stay in their enclave of nice offices with a window view all day long, they are out of sight, out of mind. They must connect with employees during and outside of training events to promote a culture, throughout the organization, that values employee training and development. That way, employees would be eager to attend training courses, rather than feeling that they are compelled to attend. Supervisors and managers would gladly schedule subordinates' work to provide for training participation, instead of doing the minimal just to comply with policies. In other words, organizational support relates to dollars and cents, as well as core values and culture.

To present a convincing business case in favor of the training program, hard and soft data are needed (Phillips, 1996a, 20). Business analysis techniques can be used to evaluate financial benefits. It is best to translate qualitative data into numerical values.

12.2 IDENTIFYING TRAINING COSTS

Training costs typically fall into the following categories:

- Compensation
- Costs of materials and equipment
- Facility and network usage
- Other expenses associated with the training program

12.2.1 Compensation

Trainees' compensation in relation to the time taken away from regular duties to participate in training is one component of training costs; so is the time of trainers, whether they are full-time or incidental trainers, for the design, development, and delivery of the training program. Do not forget the time for data collection during the needs assessment, course evaluation, and program validation processes, when other participants besides the trainees and trainers may be involved. If an organization contracts with external training consultants, professional fees are due. In addition, part of the training costs may include compensation of support personnel and any replacement or temporary staff engaged during the trainees' absence from their normal work. Compensation includes staff salaries and benefits. Benefits can be estimated as a percentage of salaries and probably range between 35% and 45%, depending on the organization and job category.

12.2.2 Costs of Materials and Equipment

Materials and equipment include training aids, media, supplies, and other apparatus. Most of the time, the costs are treated as expenses in the year in which they are incurred. Training programs that include virtual training can entail significant upfront and ongoing costs related to system evaluation, hardware purchase, software development, periodic updates, and maintenance. If substantial acquisition cost is involved, the cost may be capitalized. For instance, the cost of a simulator is $240,000 and it has a useful life of three years. You may want to treat the annual cost as $80,000, using the method of straight-line depreciation. This recommendation is solely for the purpose of estimating training costs and unrelated to how the transaction might be handled in an organization's record for financial and tax purposes. Also, course materials, equipment, and delivery platforms can become obsolete within a few years; therefore, the cost should be spread over a period of no more than three years.

12.2.3 Facility and Network Usage

A facility that belongs to the organization or meeting space that is rented is needed for in-person training. Use of the organization's facility has an opportunity cost.

Gaining Organizational Support

An organization may have an established amount of cost per square foot, including maintenance and utilities, for the use of its premises. The cost estimate can be based on square footage and time:

Cost = $X per square foot per month * Y square feet * Z months' training per year

Similarly, if the training uses hardware and software that are part of the organization's general communication and data network, the opportunity costs for such use should be estimated.

12.2.4 Other Expenses Associated with the Training Program

Travel expenses cover those related to attendance at training events for trainees, trainers, and support staff, plus other travel pertaining to the training program, such as a meeting between trainers and vendors in the procurement process. Other items include but are not limited to the costs of data collection and analysis and miscellaneous expenses at training events such as costs of refreshments.

12.3 RECOGNIZING TRAINING BENEFITS

As discussed in Chapter 11, separating training effects from confounding factors is not easy but should be done whenever possible. Improved employee performance subsequent to training may be found in better performance indicators that revolve around these dimensions:

- Productivity
- Quality
- Safety

12.3.1 Productivity

Productivity is the index of output relative to input, or results achieved relative to resources consumed. Metrics of productivity depend on the nature of the business and the job. Increased productivity may be reflected in more units of a product or service produced per unit of time, labor cost, or material cost. In an assembly line, the same employees may turn out an extra 10% of the finished product as they become more skillful after training. The contribution to profits of the additional units of finished product is the monetary benefit of productivity improvement. When the ratio of output to input increases, productivity is higher, assuming that quality is not compromised.

12.3.2 Quality

Quality means that a product or service meets expressed or implied needs and is free of deficiencies (American Society for Quality, 2013). Measurement is contingent on the type of product or service, such as the number of defective products or the response time of a service call. A widely used metric that is applicable to most businesses is customer satisfaction. While customer satisfaction is subjective, it is likely to determine whether a customer will return or refer a friend. Data on customer satisfaction is often collected through surveys that use rating scales. A more favorable rating posttraining compared with pretraining is regarded as an indication of higher quality due to training. A surrogate measure is customer retention. If the number of repeat customers increases from 500 in the previous year to 525 in the current year and the average repeat order brings in sales of $400, the contribution to profit of the additional revenues of $10,000 is the monetary benefit of quality improvement.

Another aspect of improved quality is reduced quality waste. Quality waste is the resource needed to fix defects in output (James, 1989, 3). Direct costs are incurred in correcting or replacing the defective output; indirect costs are the loss of management time, sales opportunity, and business reputation. Decreased quality waste and increased productivity are two sides of the same coin.

12.3.3 SAFETY

Also linked to productivity is a safe and healthful work environment. Effective training may be demonstrated by decreased incident rates of employee injury, property damage, or regulatory citation. Higher employee morale, like customer satisfaction, is subjective and is gauged using satisfaction surveys. Some of the surrogate measures are employee retention rate and absenteeism. For instance, in a department with 100 employees, the retention rate has increased to 95% from 90% in the year before the training program was instituted. Assuming that no confounding factors are present, such as a change in management, savings that have ensued from training would be the costs of hiring five employees and orientation to get them up to speed.

12.4 CALCULATING COST–BENEFIT OR BENEFIT–COST RATIO

Essentially cost–benefit analysis compares the inputs of the training process with the outputs. A cost–benefit ratio weighs the costs of training against the measurable results that it produces. Conversely a benefit–cost ratio shows the benefits derived from each dollar of training cost. Simply divide the monetary value of the benefits by the monetary value of the costs. Mathematically the cost–benefit and benefit–cost ratios are reciprocals of each other but the benefit–cost ratio may be easier for everyone to understand and interpret—the higher the benefit–cost ratio, the more desirable the program. For example, decreasing production time by 2,000 hours at a labor rate (salary and benefits) of $30 per hour translates into saving of $60,000 in labor cost. If the training program costs are $15,000, the benefit-cost ratio is 4, meaning that $4 in labor cost are saved for every training dollar.

Technically, the monetary values used to calculate cost–benefit or benefit–cost ratio should take into account the time value of money. An amount of $60,000 saved or spent now is worth more than the same amount saved or spent five years later. Since most benefits of training are realized in the first year, a conservative approach is to use only the first year's figures in the computation (Phillips, 1996b, 43).

Gaining Organizational Support

12.5 ANALYZING ROI

Another way to demonstrate training benefit is to calculate the ROI. More and more, organizations are examining the ROI of every program. Various methods are used to determine the ROI depending on the organization's financial structure, accounting practices, and business considerations. No matter what the criteria are, the training program must meet or exceed them. Several convenient methods that do not require in-depth knowledge of financial analysis are described below.

12.5.1 SINGLE-PERIOD ROI

Using the conservative approach as in the cost–benefit analysis and considering only the first year's investment and return, divide the net benefit by the investment to obtain the ROI. The net benefit is the total benefit minus the total cost. Needless to say, organizations favor a high ROI. Using the example in Section 12.4, the net benefit is the saving in labor cost less the training cost, or $60,000 − $15,000 = $45,000. The ROI is $45,000/$15,000 = 3, or 300%.

12.5.2 MULTIPLE-PERIOD ROI

When the costs associated with implementation or maintenance of a training program are high, it is appropriate to consider their impact in more than one year. This situation may occur when new or renovated training facilities are contemplated or when considerable cost is necessary for new course development.

12.5.2.1 Payback Period

Payback period is a simple way to assess benefits that continue over multiple years, although it is often employed to evaluate risk, not return. This analysis ignores the time value of money. It considers the period of time needed to recoup the nominal amount of the capital expenditure. It asks the question, "How long does it take to 'get back' the money invested?" Short payback periods are preferred. To find out the payback period, divide the training investment by the annual net cash flow. The annual net cash flow is the benefits minus expenses each year.

Assume that the initial investment in a training program is $100,000, the annual maintenance cost is $5,000, and the benefit is reduced work-related injuries that result in decreased workers' compensation premium and regulatory fines of $55,000 annually.

Investment	$100,000
Annual Maintenance	$5,000
Annual Savings	$55,000
Annual Net Cash Flow	$55,000 − $5,000 = $50,000
Payback Period	$100,000/$50,000 = 2 years

If annual net cash flow is constant,
Payback Period = Investment/Annual Net Cash Flow

12.5.2.2 Net Present Value (NPV)

A high NPV represents a high ROI. This method takes into account the time value of money and asks the question, "What are the net benefits in today's dollars?" NPV attempts to adjust for the effect of inflation by using a discount rate to determine the present values of future savings and costs. The challenge is to decide an appropriate discount rate, which may be subjective. The financial policy of an organization may establish a discount rate to be used for discounted cash flow analysis. The NPV can be obtained by calculating the difference between present values of savings and costs or using the "NPV" function of spreadsheet applications. Results would differ whether you assume cash flows to occur on the first day or last day of each year.

Using the previous example, suppose the training investment has a useful life of three years and the discount rate is 10%. This computation assumes that the investment and saving in the first year occur at the same time.

Investment	$100,000
Useful Life	3 years
Discount Rate	10%
Annual Maintenance	$5,000
Annual Savings	$55,000
Net Cash Flow Year 1	– ($100,000 + $5,000) + $55,000 = –$50,000
Net Cash Flow Year 2	$55,000 – $5,000 = $50,000
Net Cash Flow Year 3	$55,000 – $5,000 = $50,000
NPV	$-50,000/(1+0.1) + 50,000/(1+0.1)^2 + 50,000/(1+0.1)^3$ = $33,434

A more practical and conservative approach is to assume that the investment is paid upfront while no saving is realized until a year later.

Investment in Year 0	$100,000
Net Cash Flow Year 1	$55,000 – $5,000 = $50,000
Net Cash Flow Year 2	$55,000 – $5,000 = $50,000
Net Cash Flow Year 3	$55,000 – $5,000 = $50,000
NPV	$50,000/(1+0.1) + 50,000/(1+0.1)^2 + 50,000/(1+0.1)^3$ – 100,000 = $24,343

$$NPV = \Sigma(R_n)/(1+k)^n - I,$$

where,
 R_n = Net Cash Flow at Time n,
 k = Discount Rate,
 n = Time Period of the Net Cash Flow,
 I = Investment

Gaining Organizational Support

12.5.2.3 Internal Rate of Return (IRR)

The IRR is the discount rate that results in a zero NPV. In other words, the present value of all future net cash flow equals the investment that is paid out initially. The calculations are complex but the "IRR" function of spreadsheet applications will produce the IRR.

Organizations naturally prefer a high IRR. Note, however, that IRR and NPV may produce different results if they are used to compare programs. The reason is because IRR assumes that the net cash flow is reinvested at the same rate as the IRR, which may not be realistic and is biased against programs that have high initial investment. To address this issue, some organizations use the modified internal rate of return (MIRR). The MIRR accounts separately for the interest rate at which cash can be reinvested and for the cost of capital. Spreadsheet applications have an "MIRR" function.

Neither IRR nor MIRR should be used to compare programs of quite different sizes. A program that requires a high initial investment may have a lower IRR but a higher NPV when compared with another program that requires a lower initial investment.

An organization may have set a hurdle rate, which is the minimum acceptable rate of return (MARR) that is agreeable to the organization before starting a project, given its risk and the opportunity cost of foregoing other projects. The training program must have a rate of return that is at or above the hurdle rate.

12.6 PINPOINTING THE "HOT BUTTON"

Benefit–cost ratio and ROI analyses, along with the data on training effectiveness that you acquired during program validation, provide the bases for building the business case for your training program. You must present the case to the stakeholders to convince them to support the training program. To do this successfully, you must also understand that different stakeholders have different focus.

12.6.1 Governing Board and Senior Management

Board members and senior executives are interested in improving overall organizational performance in the most cost-effective manner. They usually want to see solid data and ROI, hence the importance of using hard data, having baseline comparisons, and being able to explain the bases of your analyses (Phillips, 1996a, 24). Although they may not be experts in statistics, they need to know that your analyses and conclusions are valid and reliable (Phillips, 1996b, 46–47). There may be exceptions to the preference of hard data. Gordon (2007, under "Deloitte & Touche") reports on organizations that rely heavily on trainee reaction surveys and may look at the ROI only when a training course is new and very expensive. Overall, even though positive reactions in surveys do not always gain organizational support, negative reactions often have an adverse effect on the training program.

From a philosophical standpoint, you want to convince the stakeholders that an organizational infrastructure delivering high-quality service must have well-trained professionals (James, 1989, 14). Since these stakeholders like to see how the training program supports organizational goals, presenting them with an impact map would be helpful (Innovative Learning Group, Inc., 2007, 1). The impact map tabulates skills and knowledge, critical tasks, key job results, and organizational goals so that the linkages between the skills and knowledge taught and organizational goals are clearly seen.

Marketing experts suggest that to eliminate a psychological barrier, the word "cost" or "expense" should be replaced by the word "investment" when presenting proposals. Training is indeed an investment as it has benefits that accrue over time. Some organizations invest for the long haul. They are prepared to forego short-term profits to achieve long-term success. Others want to see immediate returns. As discussed in Sections 12.4 and 12.5, there is more than one way to assess the benefits of a training program and different techniques may yield different results when comparing competing programs. You will find it advantageous to be familiar with organizational priorities so that you know which suitable tool to use. For example, an organization that is willing to wait for long-term results may support a training program that requires a high initial investment if it also has a high NPV, provided that the necessary funds are available. An example is a program that converts many in-person training courses to virtual training modules. To make it more palatable for funding purposes, develop a scalable program to reduce the initial investment. Start with less sophisticated courses and use third-party software packages when they can fulfill your training needs satisfactorily. Use a hosted web conferencing service to keep upfront investment low and retain the option of migrating to a managed service later. These strategies also have the advantage of allowing for improvement of the program after experience is gained as modules are rolled out gradually.

12.6.2 Middle Managers and Frontline Supervisors

Middle managers and frontline supervisors are primarily interested in the operations of their departments or business units. Although they would like their subordinates to have the knowledge and skills to do a good job, they are concerned about schedule conflicts and staff shortage when their subordinates leave regular work to attend training. Some may even see enhanced skills of subordinates as increased threats that these employees will leave for another job elsewhere.

Show positive training results from comparable departments, such as increased efficiency leading to higher productivity and profits. This approach works best if the stakeholders are accountable for profit centers. Recruit someone among their peers that could be a champion of your "cause." If possible, a face-to-face meeting is effective to secure buy-in.

Address stakeholders' concerns along these lines:

- What are the tangible and intangible benefits of the training program to their department?
- What are the training costs to their department, if these are charged back to individual departments?

- How much time would training take and how will this impact their service to internal or external clients?
- What are the costs of doing nothing?

Your goal is to reach an agreement regarding the training needs and the deliverables, documented in writing.

12.6.3 Employees

Motivated employees would probably like to participate in more training to acquire skills that help them advance in their career. They do not need much persuasion about the benefits of a training program. Still, they must be convinced that the program offered to them meets their needs including practical content, dynamic delivery, and convenient schedule. If they have provided suggestions for improvement of the training program, implement their ideas as far as reasonable and practicable. It is good to let them know why when suggestions are not adopted.

Those who are less motivated may consider training as extra work. New topics and methods of delivery, such as mobile learning, may stimulate their interest. Data from trainee reaction surveys may demonstrate how training had benefited trainees not only professionally but also personally. Such data should be shared with other employees. Peer testimonials are powerful.

Regardless of the level of motivation, the training program competes with myriads of other priorities for potential trainees' time and attention. Convenient scheduling and innovative delivery methods can help capture interest. Changing the delivery mode of existing courses can give them new appeal. A refresher course should have content and instructional strategies that are different from a basic course on the same topic. Some organizations make trainees take exactly the same courses year after year on topics that require annual training. Such repetition can only add to boredom. The fact that certain mandatory content must always be included does not mean that it should always be delivered in the same way.

12.7 PRESENTING THE BUSINESS CASE

Your presentation of the business case may be oral or written or a combination. The format probably varies depending on the group of stakeholders. Use the same presentation skills as you would use for presenting other business proposals.

- Research and plan for the target audience.
- Focus on your objectives for that audience.
- "Speak the language" of and dress appropriately for your audience.
- Describe what is in it for them.
- Use both hard and soft data.
- Explain statistics with easy-to-read charts and state the conclusions.
- Add emotional appeal.
- Be concise but be sure to address potential questions and objections.
- Supplement an oral presentation with a written summary such as a fact sheet.
- Include an executive summary in a written proposal.

12.8 SUMMARY

A training program competes with many other programs for organizational resources. It must demonstrate value, quantified in terms of a favorable benefit–cost ratio and high ROI. Various groups of stakeholders differ in their primary focus. The business case must be presented to convince members of each group of the value to them to gain commitment and support at all levels of the organization.

REFERENCES

American Society for Quality. 2013. "Quality Glossary—Q." Accessed January 29. http://asq.org/glossary/q.html.

Gordon, J. 2007. "Eye on ROI?" *Training,* May, 43–45.

Innovative Learning Group, Inc. 2007. "Impact Mapping White Paper." White Paper. Innovative Learning Group, Inc. http://www.innovativelg.com/content/secure/viewpdf.aspx?f = ILG_Impact_Mapping_White_Paper.pdf.

James, B. C. 1989. *Quality Management for Health Care Delivery*. Chicago: The Hospital Research and Educational Trust of the American Hospital Association.

Phillips, J. J. 1996a. "How much is the training worth?" *Training and Development* 50 (4): 20–24.

Phillips, J. J. 1996b. "ROI: The search for best practices." *Training and Development* 50 (2): 42–47.

13 Testing with Validity and Reliability

13.1 THE "WEIGHT OF EVIDENCE"

As explained in Chapter 8, testing and assessment are essential components of a training plan. They are an indicator of how much the trainees have learned. They provide evidence of an increase in knowledge, improvement in skill, or change in behavior as a result of the training. Unless this evidence is valid and reliable, however, it is worthless because it would not confirm achievement of the learning objectives. Validity and reliability depend on the testing methods used, construction of test items, and passing criteria. Chapter 8 discusses matching various testing methods to learning objectives. This chapter describes how to develop the tests and determine cut scores, or passing scores.

13.2 VALIDITY AND RELIABILITY

A test is valid when the test scores represent what trainees have learned based on what the training is set out to accomplish. If the training is to teach trainees how to operate a forklift but the test measures how well they drive a truck, the test is not valid.

A test blueprint, or test specification, is an essential tool in establishing test content validity. The blueprint lists the topics in the training course and their desired learning levels. The topics represent the knowledge and skills required to perform a task satisfactorily. The knowledge and skills are grouped by the various levels of learning. The percentage of questions under each learning level is specified so that the proper weight is given to each level depending on the learning objectives. Test questions are developed using the specifications in the blueprint and following best practices for test construction. Just as a lesson plan is the road map for developing the training course, the test blueprint is the road map for developing the tests. Any testing method can be evaluated for validity by comparing the test with the blueprint specifications and asking questions such as these:

- Do the test items cover the topics specified?
- Are the test items appropriate for the learning levels expected?
- Do the test items measure what they are supposed to measure?
- Are the test items properly constructed and free from ambiguity?

A test must also be reliable. That means results are repeatable in that similar scores will be achieved by individuals with similar level of knowledge and skills. A trainee that has scored 75% in one multiple-choice test would have approximately the same score in another multiple-choice test administered close in time and with a different set of questions based on the same test blueprint. The premise is that the test items are properly constructed. If not, a person taking the test may be able to repeatedly guess the correct answers; then the test would have reliability but the results are invalid. Another caution is that when evaluating reliability, external factors and trainee characteristics that may affect test results should be considered. Examples are environmental distraction, trainee's test anxiety, and trainee's familiarity with the format of the test.

With the rise in virtual training, computerized tests deserve special mention. Computerized tests come in many designs, from the linear fixed-length test (LFT) to the computerized adaptive test (CAT) (Jodoin et al., 2006, 204–205). In LFT, separate tests that are closely matched for content and item statistics are randomly assigned to different trainees. In CAT, software is programmed to score each response and select the next question based on the trainee's estimated level of knowledge or skill; the test is stopped when a predetermined statistical rule is met. Jodoin et al. (2006, 217) find that computerized tests produce accurate ability estimates and decision consistency for several credentialing examinations. Green et al. (1984, 352–353) point out that for CATs, the standard error of measurement of each score may be more meaningful than the traditional reliability index and, under certain conditions, reliability must be evaluated empirically. With regard to validity, when a CAT is used as an alternative to or replacement for a current paper-and-pencil test, equivalence of the two formats should be verified, possibly with other measures, before the CAT is considered valid (Green et al., 1984, 353). When a CAT is new, its validity should be established based on the test blueprint.

Regardless of the test format, using a test blueprint helps to ensure validity and reliability of the test and coverage of learning objectives. An additional step that provides further assurance is to have the test reviewed by subject-matter experts or run a pilot test.

13.3 FIXED-CHOICE QUESTIONS

Formats of fixed-choice questions include multiple choice, multiple select, true/false, matching, and ordering. They can be used for learning objectives at the knowledge, comprehension, application, and analysis levels. Each question usually does not require much time to answer, so a test can comprise many questions to cover a lot of materials. Scoring is objective and reliability can be evaluated. Item analysis of the answers submitted serves as a diagnostic tool to find out if and what supplemental training may be needed. The hard part is in constructing the questions, which can be time-consuming. Knowledge and application of the guidelines for creating such questions is necessary to ensure that they are clear and fair. For example, a common requirement for fixed-choice questions is that each question should only contain one idea. Double-barreled questions are to be avoided as a different answer choice may be correct for different parts of the question. Split the question into two as in the following example:

- Double-barreled true/false question:
  In the United States, federal and state laws prescribe licensing requirements of physicians.

- Split questions:
  In the United States, federal laws prescribe licensing requirements of physicians.
  In the United States, state laws prescribe licensing requirements of physicians.

13.3.1 Multiple Choice

Multiple-choice questions have an advantage over true/false questions in that the answers are less influenced by guessing. Multiple choice also avoids the absolute judgment in true/false questions. A variation of multiple choice is multiple select, where correct answers are given in more than one option.

A multiple-choice question has three parts:

- *Stem:* The statement introducing the question.
- *Key:* The correct answer.
- *Distractors or foils:* Wrong answers.

The following are some guidelines for construction of multiple-choice questions (Flateby, 2013, 14–17; USF College of Public Health, 2013, 11–15):

- Avoid long questions—they take time to read and the score may be affected by a trainee's reading ability.
- Address one idea per question and ensure that one question would not provide a hint for another question.
- Do not give away the answer by words in the stem that include or point to the correct answer.
- Write a complete statement or question in the stem or place the blank in a completion question at the end (for example, "A complete defense to a defamation suit is _____").
- Phrase the stem in the positive form; at a minimum, highlight or underline negative words (for example, "not" or "except").
- Write the stem to include words that are repeated at the beginning of all the options in the question.
- Use letter designations, not numbers, for the options and place the options in logical or numerical order (for example, 25%, 50%, 75%, 100%); they should not overlap.
- Make all options in a question about the same length, with the same type of content, and grammatically correct when read with the stem; distractors should be plausible.
- Assign only one correct option to a question; avoid options that state "never," "always," "all of the above," "none of the above," or are based on opinions.
- Change the positions of the correct options within a set of questions so that each possible position is used about as often as any other position.
- Consider placing easy items at the beginning of the test to help trainees build confidence when taking the test.

13.3.2 True/False

True/false questions are not affected by trainees' reading ability as much as multiple-choice questions but it is difficult to write questions that are unambiguous. There is a higher chance of guessing the right answer. False statements do not indicate whether the trainees know the correct answer, unless the modified true/false with correction format is used, whereby the trainee is required to correct a false statement.

The following are some guidelines for construction of true/false questions (USF College of Public Health, 2013, 16):

- Write concise statements that are absolutely true or false with no ambiguity; do not add extraneous material.
- Address one idea per question and ensure that one question would not provide a hint for another question.
- Phrase the statement in the positive form; avoid absolute terms such as "never" or "always."
- Include the same number of true and false questions and make the statements about the same length.
- Randomize the sequence of the true and false questions so they do not appear in any particular pattern.

13.3.3 Matching

Matching and ordering questions are similar. They are suitable for testing knowledge of how certain concepts or events are associated. They are hard to construct because of the need to find a set of questions and options for each test item.

The following are some guidelines for construction of matching questions (USF College of Public Health, 2013, 19–20):

- Explain the basis for the matching before listing the questions and options.
- Use numbers for the questions and letters for the options.
- Make more options available than questions; require that an option be used only once.
- Write questions (and options) that are homogeneous in content and composition.
- Use longer phrases for questions and shorter phrases for options.
- Avoid words such as "a" or "an" when they can point to the correct option.

13.4 OPEN-ENDED QUESTIONS

Open-ended questions include fill-in-the-blank, short answer, essay, case study, and situational judgment testing. They can measure higher levels of learning and are easy to create, but scoring is more subjective than fixed-choice questions. Administration and scoring of some open-ended questions also take longer, so the test would be more focused on specific content areas of the training.

13.4.1 Fill-in-the-Blank and Short Answer

These types of questions require recall of the course content. They reduce guessing and do not take much time to complete, although scoring may take longer than fixed-choice questions. Care must be taken to phrase the question in such a way that there is only one correct answer, unless multiple answers are acceptable.

The following are some guidelines for construction of fill-in-the-blank and short-answer questions (USF College of Public Health, 2013, 22):

- Write questions that require one or a few key words in the answer.
- Place the blank space at the end of the question statement.
- Avoid words such as "a" or "an" when they can hint at the correct answer.
- Provide the unit for a numerical answer.

13.4.2 Essay or Oral Explanation

Essay questions can be used for many levels of learning. They are easy to write but time-consuming to answer and score unless the oral explanation alternative is used. They may not cover a lot of the course content. They promote creativity. Consequently it is hard to establish reliability of the test. Such an issue may be mitigated by having more questions that limit responses to shorter answers, rather than fewer questions that require long answers. Shorter answers also alleviate bias against trainees that do not have good written or oral communication skills. Trainees taking the test should be informed in advance whether writing or speaking ability, as the case may be, is scored along with content.

The following are some guidelines for construction and scoring of essay questions (Flateby, 2013, 5–6; USF College of Public Health, 2013, 24):

- State a question clearly to indicate the scope and focus of the desired answer.
- Specify how long the answer should be in number of words or pages.
- Develop a model answer for each question, with a weight for each element.
- Determine score deduction for substantial deviation from the model answer.
- Score all answers to the same question before moving on to the next question.

13.5 CASE STUDY AND SITUATIONAL JUDGMENT

Tests on case study and situational judgment can be incorporated into fixed-choice questions, open-ended questions, or performance demonstration in both in-person and virtual training. A scenario is described, sometimes with a drawing, graph, or video, and the answer to an accompanying question is found or deduced from the information given. Several questions may be based on one scenario if it is complex. This testing method enables fixed-choice questions to be used for the higher levels of learning. Alternatively, a scenario can constitute the background of a performance demonstration. For example, role-playing in communication skills training may be based on a scenario in which an organization's spokesperson faces hostile citizens in a community due to an environmental disaster caused by the organization.

In addition to creating valid and reliable questions that pertain to the scenario, you should ensure that the scenario and associated materials are clearly described and understood by the trainees that are tested. Graphics or videos should have high quality. Objects that provide answers to the questions should not be buried like a needle in a haystack and, if displayed in a video, should appear on the screen long enough for trainees to see.

13.6 PERFORMANCE ASSESSMENT

Performance assessment can be in the form of demonstration, simulation, role-playing, and on-the-job observation to test learning at the higher levels. The assessment may be done in-person or virtually, at the training event or later. As always, the test should have validity based on the test blueprint. Since scoring is by observation, care must be taken to ensure reliability.

If the test is administered at the training event, you are likely to be the rater. You may want to consider allowing trainees a practice session before they are rated so that the rating is not affected by the testing environment (Mohr et al., 2000, 324–325). This may be necessary when they must use unfamiliar technology that is unrelated to the topics taught. An example is a test of skills in troubleshooting equipment carried out using a simulator. If the test is conducted later, raters may be one or more other persons, such as supervisors, peers, or the trainees themselves. The rating should be done using a standard checklist of task performance criteria, which would be derived from the task analysis. Rating scales are available in various formats, for example, the Likert 5-point or a similar scale, or simply pass/fail or yes/no to indicate whether a performance criterion is met. Several rating scales can be combined into a composite scale. The referents used in a scale should be clearly defined. All raters should be reminded to put aside personal prejudices and preferences and use only the established criteria in their ratings. They should understand how the rating scale should be used, potential rating biases, and ways to avoid them (Pedhazur and Schmelkin, 1991, 121). A frequent bias is the halo effect, when a high rating is based on positive impressions rather than the specific aspect to be rated. This may occur, for instance, when a rater gives a trainee a high rating because the trainee is a good team player, not because the trainee has demonstrated the computer skill that is being rated. The reverse halo effect can also happen, when a low rating is based on a negative experience unrelated to the skill in question. Other concerns are leniency and severity errors due to some raters giving consistently high or low ratings and the error of central tendency when raters tend to stay around the middle of the scale.

13.7 CUT SCORE

A meticulously developed multiple-choice test is given to a group of trainees. The range of scores is between 40% and 74%. Have all the trainees failed? Have they all passed? Have some failed and some passed? Of course, it depends on the cut score, which should have been notified to the trainees at the beginning of the training, but how was the cut score determined?

A basic requirement of criterion-referenced testing is that the cut score must be objectively set to ensure that all trainees taking the test have an equal chance of passing and are not affected by other trainees' scores (Le, 2009). The cut score represents the minimum level of desired result that a trainee must attain. The cut score should not be chosen arbitrarily or based on what percentage of trainees should pass.

For example, in an accredited professional certification examination, the test content and cut score are determined using psychometric procedures in a three-step process. First, a panel of experts in the profession convenes to delineate the professional role of a certificant and create a blueprint for the examination. Second, test questions based on the blueprint are written by a group of subject-matter experts. Third, a panel evaluates the examination and panel members' ratings are used to determine the cut score that represents the minimum competency required of a certificant.

For most training, and especially if you are an incidental trainer, you probably will not have the resources to convene panels in all the three steps. You and any co-trainer you may have would determine the required competency during the needs assessment and develop the learning objectives and tests accordingly. Instead of assembling a formal panel, you can run a pilot test with several subject-matter experts to set the cut score and document how the cut score is arrived at. This process should apply to oral and written tests and performance assessment.

13.8 SUMMARY

Learning should be verified by testing and assessment but a test is only valuable if it is valid and reliable. Proper question construction and scoring techniques go a long way to achieve validity and reliability. For performance that is observed and rated, the rating process presents additional issues. Some of the techniques discussed in the next chapter for designing survey questions are useful for creating rating scales for performance assessment.

REFERENCES

Flateby, T. L. 2013. "A Guide for Writing and Improving Achievement Tests." University of South Florida. Accessed April 21. http://www.acad.usf.edu/Office/IE/Resources/Guide.pdf.

Green, B. F., R. D. Bock, L. G. Humphreys, R. L. Linn, and M. D. Reckase. 1984. "Technical guidelines for assessing computerized adaptive tests." *Journal of Educational Measurement* 21 (4): 347–360. doi:10.2307/1434586.

Jodoin, M. G., A. Zenisky, and R. K. Hambleton. 2006. "Comparison of the psychometric properties of several computer-based test designs for credentialing exams with multiple purposes." *Applied Measurement in Education* 19 (3): 203–220.

Le, S. 2009. "How is a Passing Score Set?" *Board of Certified Safety Professionals BCSP Blog*. http://www.bcsp.org/BlogRetrieve.aspx?PostID = 87823&A = SearchResul.

Mohr, C. G., S. Field, and G. Frank. 2000. "Virtual reality: Is it for you?" In *The ASTD Handbook of Training Design and Delivery: A Comprehensive Guide to Creating and Delivering Training Programs—Instructor-Led, Computer-Based, or Self-Directed*, edited by G. M. Piskurich, P. Beckschi, and B. Hall. New York: McGraw-Hill, pp. 321–341.

Pedhazur, E. J., and L. P. Schmelkin. 1991. *Measurement, Design, and Analysis: An Integrated Approach*. Hillsdale: Lawrence Erlbaum Associates.

USF College of Public Health. 2013. "Assess Students." Office of Educational Technology and Assessment, USF College of Public Health, University of South Florida. Accessed April 21. http://eta.health.usf.edu/PTO/module3/unit6/booklet.pdf.

14 Conducting Meaningful Surveys

14.1 WHY YOU WANT TO KNOW ABOUT SURVEY DESIGN

You are a subject-matter expert and incidental trainer. Why would you want to know about conducting surveys, which is what researchers do? Surveys can serve as useful tools that help you gather necessary data during several processes in a training program:

- Needs assessment
- Testing and assessment
- Course evaluation
- Program validation

When you perform a training needs assessment to determine the gap between desired and actual employee performance, you can solicit input from employees and their supervisors through surveys. After the training, one way to assess trainees' transfer of learning to on-the-job application is by collecting feedback from stakeholders including customers. Reaction surveys are used to evaluate individual training courses and the whole training program. Although you can obtain information informally, solid data and statistics are much more powerful, especially in assessing program effectiveness and garnering organizational support (Phillips, 1996, 46).

Surveys involve sampling, data collection, and data processing. Sampling and data processing methodologies are beyond the scope of this book. The focus of this chapter is on data collection and how to maximize data quality and minimize bias so as to produce meaningful information. The fact that most of your surveys are probably carried out within an organization, as opposed to the general population, does not negate the need to employ strategies that would enhance the quality of your data. Employees do not automatically respond to a survey or tell you their true opinions. Poor data quality can lead you on the wrong path in your attempt to provide effective training.

14.2 WHAT AFFECTS DATA QUALITY

Surveys collect data by asking respondents a set of questions. The choices of survey modes continue to expand with advances in technology, and you will need to pick those most suitable for your purposes. Measuring the wrong metric will produce invalid data. As discussed in Chapters 10 and 13, the focal point of a question and manner in which questions are asked and sequenced can influence the responses.

Another source of bias is who responds to the survey and who does not. A low response rate in a survey may result in skewed data if the respondents and nonrespondents have different characteristics.

14.2.1 SURVEY MODE

One of the first decisions to be made is the data collection mode. The following are the most common:

- *Mail or e-mail survey:* The message includes the survey questionnaire and is not merely an invitation to participate that provides a hyperlink to the instrument.
- *Web survey:* The survey instrument is posted on the Internet or intranet and accessible via personal computers and mobile devices.
- *Face-to-face interview:* The interviewer and interviewee meet in person or via a web meeting.
- *Telephone interview:* The interview is conducted over the telephone with no video link.
- *Computer-assisted interviewing (CAI):* There are several variations, including computer-assisted self-administered interviewing (CASI) in which the respondent reads the questions on a screen or listens to the questions, and the randomized response technique (RRT).

The RRT is used for sensitive topics such as substance abuse to minimize evasive response bias due to respondents' reluctance to disclose true information (Lensvelt-Mulders et al., 2005; Warner, 1965). Those topics are not normally within the scope of training evaluation and RRT will not be discussed further.

A survey questionnaire can be sent to potential respondents by mail or e-mail. It is self-administered. They return the instrument after completion through the same delivery method. The mail survey is declining due to the speed, convenience, and low cost of electronic surveys. An e-mail survey is an efficient method to reach a large number of people. The cost of distribution is low and delivery is fast. Generally e-mail surveys have lower response rates than paper surveys in all populations including college students and university faculties (Shannon and Bradshaw, 2002, 189; Shih and Fan, 2009, 36). An e-mail survey is not anonymous. Even if potential respondents are informed that names will be removed for data analysis, they may not want to participate (Kittleson, 1997, under "Discussion"). Other concerns regarding e-mail delivery are discussed later in this chapter.

Web survey avoids the problems of surveys lost in the mail or resource requirements for data input. Software applications facilitate data analysis. The cost of distribution of the survey is low. The setup cost may be high if a special website is developed for a survey. Web survey has gained popularity quickly not only because of its advantages but also because limited service is available free of charge from certain web-based providers of survey solutions. This survey mode is convenient for respondents as submission of the survey instrument is automatic once it is completed. It is cost-effective for the researcher. A web survey is particularly suitable when potential respondents are within the same organization and the survey can be set up easily on the intranet.

An interview can elicit deeper information than a self-administered questionnaire but requires a lot of resources. A face-to-face interview is the most expensive and time-consuming to conduct. In a face-to-face or telephone interview and in some CAI, interaction occurs between interviewer and interviewee. This can be an advantage and a disadvantage. The interviewer can clarify questions and reduce the chance of inaccurate or missing responses, or ask follow-up questions spontaneously to probe further into an interviewee's response. However, when an interview is not standardized, responses are susceptible to interviewer effects and other biases. Self-administered surveys including CASI make it convenient for respondents to participate in the study and are less costly as no interviewer is required. CAI can eliminate out-of-range responses that are not among valid codes. Smart branching technology allows questions to be personalized based on responses to previous questions (Wright et al., 1998, 333). Respondents' attitudes toward computers, confidentiality, and privacy influence results of CASI compared with paper self-administered questionnaires (347). The absence of nonverbal cues as in face-to-face interview reduces interviewer effects but may increase nonresponses to individual questions. Making support available to respondents when needed helps to reduce nonresponse. Data backup and security must be built into any electronic survey system to prevent data loss and protect private information.

The efficacy of a survey mode is affected by respondents' acceptance of the mode for the purpose of that survey and trust in its proper administration. When a survey is accessible with an individualized password to allow for response tracking, potential respondents must be convinced of the anonymity aspect. An anonymous, self-administered survey affords privacy and confidentiality and may encourage respondents to be candid (Tourangeau and Smith, 1996, 299); otherwise embarrassment on the part of the respondent can produce dishonest answers. For example, a trainee may not want to admit in an interview that he or she has not been practicing a technique taught in a training course because he or she has forgotten how to do it, but may be comfortable in giving an honest answer in an anonymous survey. Anonymity also reduces the self-presentation effect that includes prestige bias, which is respondents' tendency to answer in a way that makes them feel good (Pedhazur and Schmelkin, 1991, 140–141). A self-administered survey can save you much time since training is your incidental responsibility.

A combination of data collection modes may compensate for weaknesses of individual modes and acquire the most complete data at affordable cost (de Leeuw, 2005, 235). A mixed-mode survey may increase the response rate (Shannon and Bradshaw, 2002, 190; Yun and Trumbo, 2000, under "Conclusion"). This approach raises issues about data integrity. The data collected may not be comparable across different modes. A unimode design with the same question layout for the web and paper survey may be inappropriate (de Leeuw, 2005, 247). In a generalized mode design, the same stimulus, rather than the same questions, is presented in each mode but research is lacking on what can be considered the same stimulus (249–250). Dillman and Smyth (2007) discuss the problem of measurement errors when telephone survey questions are used in a web survey or when these two modes are combined. That is because respondents in the web survey apply cultural rules to visual features, altering the meaning originally communicated in the oral, telephone mode. In a population that has a high degree of interest in the survey topic, Yun and Trumbo (2000, under "Conclusion") find few observable differences in the quality of responses when using mail, e-mail, and web surveys with coordinated timing.

14.2.2 Metric Validity

Survey results are meaningless if what is measured is not a valid metric for the purpose of the survey. Some training courses aim at changing attitudes. Ethics training is an example. It may incorporate transfer of knowledge in ethical theories but the expectation is that eventually trainees will achieve learning in the affective domain at the highest level—characterized by a value set—and exhibit ethical behavior. Until a situation of ethical dilemma arises, the behavior cannot be observed. A surrogate is needed to measure achievement of this learning objective. One metric might measure how an employee would feel about confronting his or her direct supervisor when the latter performs an unethical act. In other cases such as clinical training, surrogates are used due to cost, practicality, and ethical reasons. For instance, assessment of performance of a patient care task using simulation would be measuring surrogates since the performance is not real.

Much research on the selection of surrogates has been in the field of clinical practice but appropriateness of a surrogate is fundamental to the validity of all inquiries. Correlation between a surrogate and the true outcome is a necessary but not sufficient condition to establish validity of the surrogate. Consideration must be given also to whether the surrogate and the true outcome are affected in the same direction and to the same extent by the training intervention. An inappropriate surrogate may display different effects in the experimental group and the control group (Baker and Kramer, 2003). A surrogate in a simulated environment may be valid for one task but not another (Hardman et al., 2008, 595). The validity of a surrogate often has to be verified empirically.

14.2.3 Question Content

The construction of test questions in testing and assessment should be guided by a test specification, as discussed in Chapter 13. In a similar fashion, a table of specification serves as a blueprint for developing survey questions (Turocy, 2002, S-175). The table of specification lists the topics and subtopics to be included in the instrument and the question numbers. The first topic can be general information such as job title and length of experience.

Recall the discussions in Chapters 10 and 13 on question design. In particular, avoid leading and loaded questions. Questions should be concise, specific, and unambiguous so they do not mean different things to different respondents. If you ask a question about how frequently respondents attend training courses in-house, state the response options in time periods (for example, "once a month," "once a quarter") instead of qualitative descriptions (for example, "very often," "often"). The response options should cover all possibilities for that question. Use terms that are familiar to the respondents. For instance, during a needs assessment in the information technology department, a trainer may interview data entry clerks and programmers. The data entry clerks may not understand technical terms used in programming. If it is necessary to use these terms during the interview, they should be clearly explained. Avoid terms that may be misinterpreted or not understood by people from diverse cultures. Refrain from using the options of "Don't know" and "No opinion" as they may have different meanings for different people. These considerations are especially relevant when a survey is self-administered with no interviewer to help with interpretation.

In an interview, consider asking a few icebreaker questions first to build rapport with the interviewee. However, avoid interviews that require an inordinate amount of the interviewees' time. Long interviews cause interviewee and interviewer fatigue.

14.2.4 QUESTION PRESENTATION

Arranging response options for a question in a logical progression helps respondents answer quicker (Tourangeau et al., 2004, 390–391). Earlier questions can influence how respondents interpret and answer later questions, due to the cognitive processes of a respondent when contemplating an answer (Tourangeau and Rasinski, 1988). Surveys that measure attitudes are particularly susceptible to context effects, which can be created by the question sequence within an instrument or option sequence within a question. Some electronic survey instruments are designed to be dynamic in question layout and ordering. Not enough is known about their effects (Fan and Yan, 2010, 134).

When constructing questions, it is advisable to think ahead about data analysis, which would be much easier if certain data are collected in a specific format. Prescribing the format reduces the chance of receiving unusable data. For example, if you ask respondents the month and year when they started their current job, you may want the month and year to be in the two-digit and four-digit format, respectively. A smaller box for the month and a larger box for the year, plus the use of symbols (MM YYYY) instead of words ("Month" and "Year"), increase the percentage of respondents using the correct format (Christian et al., 2007, 123).

A closed-ended question written in two formats that are meant to solicit similar responses can produce different results. Suppose you are conducting a survey as part of your training program validation. The program consists of five training courses. You may ask a question about usefulness of the courses in one of two formats:

Forced choice

Have You Found Each of the Following Courses Practical or Not Practical for Your Daily Work?

	Practical	Not Practical
Course A	☐	☐
Course B	☐	☐
Course C	☐	☐
Course D	☐	☐
Course E	☐	☐

Check-all-that-apply

> **Which of the Following Courses Have You Found Practical for Your Daily Work? Please Check All That Apply.**
>
> Course A ☐
> Course B ☐
> Course C ☐
> Course D ☐
> Course E ☐

The forced-choice format is found to encourage respondents to think more deeply and select more options (Smyth et al., 2006a, 75; Smyth et al., 2008, 111).

Open-ended questions are expected to receive a variety of responses. The quality of the responses may depend on how the question is presented. Smyth et al. (2009, 336) find that response quality improves when a question is accompanied by clarifying and motivating instructions, such as emphasizing the importance of the question, and, in the case of late respondents, by increasing the size of the answer box.

Tourangeau et al. (2004) examine how respondents interpret questions based on visual cues. If the response options to a question include substantive followed by nonsubstantive options such as "Don't know" or "No opinion," respondents may take the visual midpoint as the conceptual midpoint of the scale and use it as a point of reference, unless a divider line or space is inserted to separate the two types of answers. Spacing between substantive options should be evenly distributed. Respondents consider questions placed visually close together to be related and may use the same answer even when the questions are worded in opposite directions. Christian and Dillman (2004) also find that respondent behavior is affected by visual design. When the response options comprise a rating scale, order them vertically rather than horizontally in two rows because respondents might focus on the top row. Any rating scale should have an accompanying instruction on how to use the scale. Present rating scales within a survey consistently, for example, all scales go from the low to the high ratings and vice versa. If graphic rating scales are provided, ensure that the sizes of the segments on the scale correctly reflect the relative positions of the points on the scale.

Since visual presentation of questions and response options impacts the responses, an electronic survey instrument should be designed to appear as intended across different browsers and devices. Of special note is the possible change in spacing that might result in inadvertent grouping of response options, causing respondents to pick answers from what they perceive as subgroups (Smyth et al., 2006b, 14).

Dillman and Smyth (2007) recommend the following design guidelines for web surveys:

- Articulate questions and response options clearly.
- Consider if meaning is intact before applying questions from another survey mode.

- Maintain consistency in the use of design elements such as font style.
- Select visual features that have desirable effects.
- Avoid using multimedia due to unknown impact on responses and download time.
- Validate the instrument before implementation.

Additionally, in a web survey where closed-ended questions are asked, make sure that the software does not provide a default answer, or a respondent might select it inadvertently or think that it is the desirable answer. Place a note next to a text box for an open-ended question to inform respondents of the maximum number of characters allowed (including or excluding spaces) or the fact that there is no limit, as the case may be, so they know how much they can write. There is no consensus regarding whether to present all questions in a web survey on one screen or one question per screen. Tourangeau et al. (2004, 390) suggest that respondents may infer relatedness of questions on a single screen due to visual proximity. Peytchev et al. (2006, 604) find few differences between the two designs and recommend that the choice should depend on the survey's goals.

Other suggestions with respect to question presentation include using a "human-like" conversational style in CAI to simulate dialog with a human interviewer (Peiris et al., 2000, 647) and providing respondents with the questions before an interview to shorten the interview time (Wassink et al., 2004, 64).

14.2.5 RESPONSE RATE

Response rate is important because of possible nonresponse bias. A low response rate raises the question whether the respondents are representative of the study population and whether they have similar characteristics as the nonrespondents. These factors impact data quality.

There is a general downward trend of survey responses in many countries, probably due to survey fatigue or "survey burden" (Baruch, 1999, 432; de Heer, 1999, 139; Goyder, 1986, 39). In addition, e-mail surveys may be ignored by potential respondents due to the increase of unsolicited e-mails over time (Sheehan, 2001, under "Discussion"). Response rate varies with the study purpose, survey mode, and population. In a study of health professionals, Kittleson (1997) obtained a 50% response rate. Interestingly, Baruch (1999, 433) finds that the most decline in survey response involves top management or organizational representatives. Still, employees are more likely to respond to surveys than the general population (Heberlein and Baumgartner, 1978, 459–460).

Some potential respondents may not complete the survey at all. Some respondents may not answer all questions. Good instrument design and interviewer skill in building trust can help to improve response rates. The main reasons for nonresponse are that the potential respondents have not received the survey or they do not want to complete it (Baruch, 1999, 422; de Heer, 1999, 139).

14.2.5.1 Survey Delivery

Potential respondents can only complete a survey if they have received it. Mail deliveries can be lost in transit. E-mails can be relegated to the junk e-mail box by spam filters or removed by antivirus software. Some e-mail systems automatically purge or archive e-mails after a number of days if no action is taken by the recipient. These issues may be relatively infrequent when the recipients are employees of your organization but can be more serious when they are in other organizations, as when you are performing a follow-up customer satisfaction survey. A prenotification is effective in alerting potential respondents of a forthcoming survey. Individually sent e-mails are less likely to be blocked by spam filters than a broadcast message. Personalize the invitation to participate (Barriball and While, 1999, 683; Muñoz-Leiva et al., 2010, 1049). Explain the purpose of the study, provide motivation why it is important to participate, assure recipients of confidentiality, inform them how long it would take to complete the survey, and include a date for completion. Supply clear instructions to make it easy for them to locate the survey instrument and guide them through the process of the survey (Sheehan and McMillan, 1999, 52). Let them know how they can get help if needed. Respondents that are frustrated due to technical difficulties encountered in completing the instrument are not likely to submit their responses. If the survey instrument has multiple pages, it is advisable to place the process and support instructions on each page.

14.2.5.2 Survey Completion

Lowering the costs to respondents of completing the survey will increase response rate (Heberlein and Baumgartner, 1978, 458). Costs are not necessarily monetary. They include the time to complete and return the instrument. A visually attractive layout helps to improve response rate (Ravichandran and Arendt, 2008, 64). In a mail survey, provide a self-addressed stamped envelope. Electronic submission saves the task of having to mail the instrument back. Another way to reduce costs is to increase motivation by emphasizing the importance of the study and the individual's response, sending a personal communication, and making follow-up contact. Kittleson (1997, under "Results") finds that in an e-mail survey, follow-up doubles the response rate but there is no difference whether one, two, or four reminders are sent. How soon to send reminders would depend on the complexity of the instrument and the urgency of completing the study (Muñoz-Leiva et al., 2010, 1049–1050). You may want to send the first reminder in four to seven days (Kittleson, 1997, under "Recommendations"). Consider using alternative and multiple methods of communication, such as texting and a telephone call.

Employees are likely to complete organizational surveys, especially when they feel that the organization will act on the survey data. Noncompliance does exist and may be due to reasons similar to other populations, such as lack of time, or specific to the work environment, such as lack of commitment to the organization (Rogelberg et al., 2000, 291). Employee response rate may be adversely affected by impending organizational changes and layoffs (Barriball and While, 1999, 684). Such factors may lead to nonresponse to the survey or individual questions. Some employees may miss a survey due to being away on vacation. The upward trend of part-time staffing increases the likelihood of noncompliance. Scheduling of interviews must be flexible to capture employees who work night shifts or weekends.

14.3 WHAT TO CHECK IN A PILOT TEST

Running a pilot test and making adjustments as necessary ensure that an instrument is valid and reliable for the survey mode it is to be used. Ask the following questions:

- Are the instructions easy to follow?
- Is word usage appropriate for the potential respondents?
- Are rating scales simple to interpret?
- Are the response options within a question mutually exclusive?
- Do the response options within a question cover all possible answers?
- Are there confusing questions or response options?
- Is the question sequence likely to bias responses?
- Do the responses provide the data you set out to collect?
- How long does it take to complete the survey?
- Are the features of a web survey supported by various browsers and devices?
- Is the visual appearance of a web survey consistent across browsers and devices?

In addition, check the data import and export functionalities of an electronic survey during the review process (Fan and Yan, 2010, 137).

14.4 SUMMARY

Surveys provide a valuable tool for collecting data that are useful in a training program. Careful choice of survey modes, selection of valid metrics, and design of survey instruments, together with timely communication, will contribute to high response rates and data quality.

REFERENCES

Baker, S., and B. Kramer. 2003. "A perfect correlate does not a surrogate make." *BMC Medical Research Methodology* 3 (1):16. http://www.biomedcentral.com/1471-2288/3/16.

Barriball, K. L., and A. E. While. 1999. "Non-response in survey research: A methodological discussion and development of an explanatory model." *Journal of Advanced Nursing* 30 (3): 677–686.

Baruch, Y. 1999. "Response rate in academic studies—A comparative analysis." *Human Relations* 52 (4): 421–438. doi:10.1177/001872679905200401.

Christian, L. M., and D. A. Dillman. 2004. "The influence of graphical and symbolic language manipulations on responses to self-administered questions." *Public Opinion Quarterly* 68 (1): 57–80. doi:10.1093/poq/nfh004.

Christian, L. M., D. A. Dillman, and J. D. Smyth. 2007. "Helping respondents get it right the first time: The influence of words, symbols, and graphics in web surveys." *Public Opinion Quarterly* 71 (1): 113–125. doi:10.1093/poq/nfl039.

de Heer, W. 1999. "International response trends: Results of an international survey." *Journal of Official Statistics* 15 (2): 129–142.

de Leeuw, E. D. 2005. "To mix or not to mix data collection modes in surveys." *Journal of Official Statistics* 21 (2): 233–255.

Dillman, D. A., and J. D. Smyth. 2007. "Design effects in the transition to web-based surveys." *American Journal of Preventive Medicine* 32 (5, Supplement): S90–S96. doi:10.1016/j.amepre.2007.03.008.

Fan, W., and Z. Yan. 2010. "Factors affecting response rates of the web survey: A systematic review." *Computers in Human Behavior* 26 (2): 132–139. doi:10.1016/j.chb.2009.10.015.

Goyder, J. 1986. "Surveys on surveys: Limitations and potentialities." *Public Opinion Quarterly* 50 (1): 27–41. doi:10.2307/2748968.

Hardman, J. G., I. K. Moppett, and R. P. Mahajan. 2008. "Validity, credibility, and applicability: The rise and rise of the surrogate." *British Journal of Anaesthesia* 101 (5): 595–596. doi:10.1093/bja/aen292.

Heberlein, T. A., and R. Baumgartner. 1978. "Factors affecting response rates to mailed questionnaires: A quantitative analysis of the published literature." *American Sociological Review* 43 (4): 447–462. doi:10.2307/2094771.

Kittleson, M. J. 1997. "Determining effective follow-up of e-mail surveys." *American Journal of Health Behavior* 21 (3): 193. CINAHL Plus with Full Text.

Lensvelt-Mulders, G. J. L. M., J. J. Hox, P. G. M. van der Heijden, and C. J. M. Maas. 2005. "Meta-analysis of randomized response research: Thirty-five years of validation." *Sociological Methods & Research* 33 (3): 319–348. doi:10.1177/0049124104268664.

Muñoz-Leiva, F., J. Sánchez-Fernández, F. Montoro-Ríos, and J. A. Ibáñez-Zapata. 2010. "Improving the response rate and quality in web-based surveys through the personalization and frequency of reminder mailings." *Quality and Quantity* 44 (5): 1037–1052. doi:10.1007/s11135-009-9256-5.

Pedhazur, E. J., and L. P. Schmelkin. 1991. *Measurement, Design, and Analysis: An Integrated Approach.* Hillsdale: Lawrence Erlbaum Associates.

Peiris, D. R., P. Gregor, and N. Alm. 2000. "The effects of simulating human conversational style in a computer-based interview." *Interacting with Computers* 12 (6): 635–650.

Peytchev, A., M. P. Couper, S. E. McCabe, and S. D. Crawford. 2006. "Web survey design: Paging versus scrolling." *Public Opinion Quarterly* 70 (4): 596–607. doi:10.1093/poq/nfl028.

Phillips, J. J. 1996. "ROI: The search for best practices." *Training & Development* 50 (2): 42–47.

Ravichandran, S., and S. W. Arendt. 2008. "How to increase response rates when surveying hospitality managers for curriculum-related research: Lessons from past studies and interviews with lodging professionals." *Journal of Teaching in Travel and Tourism* 8 (1): 47–71. doi:10.1080/15313220802410054.

Rogelberg, S. G., A. Luong, M. E. Sederburg, and D. S. Cristol. 2000. "Employee attitude surveys: Examining the attitudes of noncompliant employees." *Journal of Applied Psychology* 85 (2): 284–293. doi:10.1037//0021-9010.85.2.284.

Shannon, D. M., and C. C. Bradshaw. 2002. "A comparison of response rate, response time, and costs of mail and electronic surveys." *Journal of Experimental Education* 70 (2): 179–192.

Sheehan, K. B. 2001. "E-mail survey response rates: A review." *Journal of Computer-Mediated Communication* 6 (2). http://jcmc.indiana.edu/vol6/issue2/sheehan.html.

Sheehan, K. B., and S. J. McMillan. 1999. "Response variation in e-mail surveys: An exploration." *Journal of Advertising Research* 39 (4): 45–54.

Shih, T.-H., and X. Fan. 2009. "Comparing response rates in e-mail and paper surveys: A meta-analysis." *Educational Research Review* 4 (1): 26–40. doi:10.1016/j.edurev.2008.01.003.

Smyth, J. D., L. M. Christian, and D. A. Dillman. 2008. "Does 'yes or no' on the telephone mean the same as 'check-all-that-apply' on the web?" *Public Opinion Quarterly* 72 (1): 103–113. doi:10.1093/poq/nfn005.

Smyth, J. D., D. A. Dillman, L. M. Christian, and M. McBride. 2009. "Open-ended questions in web surveys: Can increasing the size of answer boxes and providing extra verbal instructions improve response quality?" *Public Opinion Quarterly* 73 (2): 325–337. doi:10.1093/poq/nfp029.

Smyth, J. D., D. A. Dillman, L. M. Christian, and M. J. Stern. 2006a. "Comparing check-all and forced-choice question formats in web surveys." *Public Opinion Quarterly* 70 (1): 66–77. doi:10.1093/poq/nfj007.

Smyth, J. D., D. A. Dillman, L. M. Christian, and M. J. Stern. 2006b. "Effects of using visual design principles to group response options in web surveys." *International Journal of Internet Science* 1 (1): 6–16.

Tourangeau, R., M. P. Couper, and F. Conrad. 2004. "Spacing, position, and order: Interpretive heuristics for visual features of survey questions." *Public Opinion Quarterly* 68 (3): 368–393. doi:10.1093/poq/nfh035.

Tourangeau, R., and K. A. Rasinski. 1988. "Cognitive processes underlying context effects in attitude measurement." *Psychological Bulletin* 103 (3): 299–314.

Tourangeau, R., and T. W. Smith. 1996. "Asking sensitive questions: The impact of data collection mode, question format, and question context." *Public Opinion Quarterly* 60 (2): 275–304. doi:10.2307/2749691.

Turocy, P. S. 2002. "Survey research in athletic training: The scientific method of development and implementation." *Journal of Athletic Training* 37 (4 Supplement): S-174–S-179.

Warner, S. L. 1965. "Randomized response: A survey technique for eliminating evasive answer bias." *Journal of the American Statistical Association* 60 (309): 63–69. doi:10.2307/2283137.

Wassink, H. L., G. E. Chapman, R. Levy-Milne, and L. Forster-Coull. 2004. "Implementing the British Columbia nutrition survey: Perspectives of interviewers and facilitators." *Canadian Journal of Dietetic Practice and Research* 65 (2): 59–64.

Wright, D. L., W. S. Aquilino, and A. J. Supple. 1998. "A comparison of computer-assisted and paper-and-pencil self-administered questionnaires in a survey on smoking, alcohol, and drug use." *Public Opinion Quarterly* 62 (3): 331–353. doi:10.2307/2749663.

Yun, G. W., and C. W. Trumbo. 2000. "Comparative response to a survey executed by post, e-mail, & web form." *Journal of Computer-Mediated Communication* 6 (1). http://jcmc.indiana.edu/vol6/issue1/yun.html.

15 Leveraging Generational Learning

15.1 FOUR GENERATIONS AT WORK

The late 20th century has witnessed for the first time in history four generations working together in the occupational environment. That is because financial necessities and longer lifespan have prompted people to remain in the work force after what used to be the normal retirement age. The Bureau of Labor Statistics (2012, 1) projects that workers in the 55-years-and-older age group in the United States will comprise 25.2% of the labor force in 2020, up from 19.5% in 2010.

Social scientists have referred to these four cohorts by different names. Other than "Baby Boomers," the terminology and age groupings vary. One classification goes along these lines (American Management Association, 2007):

- *Silents:* Born between 1925 and 1946.
- *Baby Boomers:* Born between 1946 and 1964.
- *Generation Xers:* Born between 1965 and 1980.
- *Millennials:* Born after 1980.

Individual differences abound and stereotyping should be avoided. Common characteristics of the cohorts do offer a starting point in understanding how and why some trainees may react or interact in certain ways, possibly due to their generation's history and life experiences. Some of the characteristics are described below primarily with reference to the North American population.

15.1.1 Silents

Also known as "Veterans," "Traditionals," or "Seniors," members of this generation are conformers, hence the term "Silents." Many grew up during the Great Depression and are value-driven. They submit to authority and discipline. In their view, working hard is the right thing to do. They appreciate teamwork and are loyal employees but not risk takers. They prefer consistency, stability, and security. They may be resistant to change and need to be convinced why it is necessary.

15.1.2 Baby Boomers

Baby Boomers have lived through the Watergate scandal and the civil rights movement. They do not submit to authority like the Silents and are also more adaptable to change. The Baby Boomer generation is known as the "Me Generation" since its members emphasize personal growth. They place a high priority on work, more than personal life, to succeed. They understand that peer competition is necessary for success.

15.1.3 Generation Xers

Generation X is called the MTV Generation, referring to the television channel that changed the music industry. With both parents working, they grew up as latchkey kids, in an era of economic decline and increased divorces. They tend to be independent and cynical. Having observed their workaholic parents, the Xers place emphasis on work–life balance. They work hard to find ways to do things faster so they have more time for leisure. They are adaptable, entrepreneurial, and technologically savvy.

15.1.4 Millennials

The Millennials are also named "Generation Y," "Generation Next," or "Nexters." The ethnic composition of this generation is the most diverse of the four generations and the members embrace diversity. They grew up in a global economy and the age of high technology. Their parents are protective and engaged them in activities that promoted self-esteem and prepared them for success. They value positive reinforcement. They are goal-oriented and work well in teams. Being good at multitasking, they find it hard to focus on one thing at a time.

15.2 FOUR GENERATIONS IN TRAINING

With four generations in the workplace, trainers need to consider how generational characteristics may affect the learning process, keeping in mind, once again, that generalizations do not necessarily apply to each individual and learning preferences are a function of prior experience with the subject matter (Thompson and Sheckley, 1997, 168).

15.2.1 Training the Silents

Ageists disparagingly use the expression "You can't teach an old dog new tricks" to discredit older workers' ability to learn. Scientific evidence has proved the contrary. Reasoning and comprehension are not affected adversely by older age in healthy individuals (Small et al., 1999, under "Discussion"), and older workers may benefit from the healthy worker effect (McMichael, 1976). Although decrement in cognitive tasks has been associated with increasing age, research has shown that memory training can improve performance (Cavallini et al., 2003, 253).

Older workers may take longer to digest new information or learn new technology. For example, they need more time than younger workers to become proficient in navigating the World Wide Web (Laberge and Scialfa, 2005, 300). If an organization would like to improve the computer literacy of its older workers but does not have the expertise to accommodate their special needs, nonprofit computer learning centers with specially trained instructors for the older population are available

(SeniorNet, 2013). Do not assume, however, that all members of this generation have no computer skills. According to a survey of the Pew Research Center (2012, 4, 6), as of April 2012, 53% of American adults aged 65 years or older and 34% of those aged 76 years or older use the Internet or e-mail.

Silents value learning skills that are relevant to their job but may prefer to practice in private so they do not risk "losing face" in the presence of others. Since they are not risk takers, it is not surprising that they prefer a stable learning environment free of risks (Zemke et al., 1999, under "Veterans"). You may want to establish ground rules at the beginning of the training so the Silents know what is expected and will conform. Display the rules in a conspicuous place or distribute them in a handout. The Silents' respect of authority may extend to you as the trainer. They may not tell you when they disagree with you. At the same time, they may not like learning from a young trainer, so building credibility and rapport is important. Show that you respect their work and life experience and give them the opportunity to share their stories if relevant to the training topic. Avoid jargon and get to the point (Wan, 2010, 33). Use illustrations and visual aids both during the training and in the workplace as reminders.

The training aids and physical environment should cater to the Silents. For instance, if they must use computers during the training, the monitors should be large size, with text and other items displayed in the "medium" or "large" setting on the screen; the input devices should be easy to operate (Wan, 2010, 32). Hard copies of handouts should have text in a 12-point font or larger.

Suppose you are training a group of Silents on emergency evacuation procedures. Although the floor plan and evacuation routes are posted in various places in the building, you should go ahead and distribute a handout with the map on one side and step-by-step procedures on the other side so that the trainees can have their personal copies to keep for reference when needed. Be sure that the text and graphics are clearly legible.

15.2.2 Training the Baby Boomers

Baby Boomers are also "older workers" as they are now over 40 years of age and protected in the United States by federal law, the Age Discrimination in Employment Act of 1967. They are beginning to share some of the physical issues as the previous generation, such as declining vision. Unlike the Silents, Boomers will probably not consider their trainer a superior. You must build trust by treating them as equals and sharing personal examples that relate to the training topic (Zemke et al., 1999, under "Boomers"). Practical examples also guide the Boomers in transferring knowledge to behavior. As the "Me Generation," they are interested when what they learn can help them in self-development and professional advancement. They enjoy an interactive learning experience. Instructional strategies that employ group discussions, team activities, and collaborative learning fit the Boomers well. Include problem-solving and situational judgment exercises.

If you are conducting the same emergency evacuation procedures training as mentioned earlier and your trainees are Boomers, they would be interested to hear your own experience going through an emergency situation. Perhaps on one occasion the building where you worked had a fire. All personnel had to leave the building. During the head count at the assembly area, your group found that one person was missing. Everyone was worried about this coworker. You can describe how the group tried to determine where the missing person was last seen, how this was reported to the safety officer in charge, and how eventually it was found that as soon as the person came out of the building, he went home! He was absent from the safety training due to illness just two days ago and had not taken the make-up training. He was unaware of the assembly area and head count. Sharing such experience would give you the opportunity to illustrate how the procedures are carried out in a real-life situation and why it is important for all employees to be properly trained in these procedures.

15.2.3 Training the Generation Xers

Generation Xers prefer doing things in their own time. They do not like structured schedules, so self-directed learning is suitable for them (Zemke et al., 1999, under "The Xers"). A good choice might be self-paced virtual training. They learn better with visual stimulation than reading lots of text. They want to get through the information in the easiest and quickest way. Chunking of content in e-learning and mobile learning works well for them. Help and additional resources should be available when needed. The Xers respect trainers who have demonstrated expertise and may ask many questions. Their motivation to learn comes from the realization that knowledge and skills would increase their professional worth. They enjoy the learning experience if it is fun and allows them to experiment; therefore, game activities and role-playing are good instructional strategies.

In the emergency procedures training, for example, you can let the Xers role-play in an evacuation drill. Ask the trainees to pick their own roles as safety officer, operations manager, plant manager, evacuation warden, firefighter, emergency medical technician, police officer, and so on. Let them plan, organize, and perform the drill while you observe and provide guidance as needed. Upon completion, debrief them on commendable behavior and areas for improvement. You can use this exercise as an assessment tool. If you use it for a summative assessment, be sure to inform them at the beginning of the training that this is how they will be tested, whether they will receive a group or individual score, and what the evaluation criteria are.

15.2.4 Training the Millennials

The Millennials grew up in a new world of technology. They are not only computer literate. They are accustomed to multimedia. This is the generation that would appreciate mobile learning the most for its rich content and 24/7 availability. Just like the Xers, the Millennials are interested in gaining knowledge and skills that will improve their professional worth (Zemke et al., 1999, under "Generation Next"). Making—and spending—money is important to them. They are motivated when the training will lead to a positive financial outcome. They like course materials that have visual appeal, and they are better readers than the Xers. Training that incorporates creative media, interactive exercises, and experiential learning is attractive to the Millennials. They attempt to solve problems with the same mind-set as when they play video games—through trial-and-error, not logic. They will learn from their mistakes quickly, making it desirable to provide them with systematic and frequent feedback. Simulation that enables them to practice and receive immediate feedback on errors is ideal. They also enjoy solving problems in case studies and team competition in game activities.

The best way to train this group in the emergency evacuation procedures is the virtual training delivery mode. The policies and procedures can be electronically distributed through a mobile learning portal. Video clips can dramatize one or more incidents, followed by quizzes on how procedures are carried out correctly and incorrectly as shown in the videos. Additional assessment can be done by presenting a scenario in an interactive video game and having a trainee respond to situations that arise during an emergency, such as initiating the emergency alert system, shutting down the plant, or evacuating personnel. Characteristics of the simulation would depend largely on the training budget but the video game should have the capability of immediate feedback on the course of action taken by the trainee.

15.2.5 Training Four Generations Together

When the four generations are trained separately, instructional strategies and course materials can be developed for their respective learning preferences. Training can be delivered in the classroom for some and virtually for others. For many general training topics, chances are that the four generations in the same workplace will attend in-person training together. This poses challenges and opportunities for trainers.

15.2.5.1 A Challenge and an Opportunity

On the one hand, you must balance and cater to the needs and expectations of a diverse group. On the other hand, the diversity can enrich the learning experience of all trainees. It is important to acknowledge the characteristics of each generation while avoiding stereotyping divergent individuals (Johnson and Romanello, 2005, 214). With that understanding, you can use events and icons to which members of the various generations can relate (Zemke et al., 1999, under "All in One Room"). As examples, the Great Depression, the Vietnam War, the Iran hostage crisis, and the September 11 terrorist attacks were significant events for each of the generations.

Trainees from the Silents generation may be able to offer insight as to how work has evolved over time. For instance, their description of how a task used to be done "back in those days" when they were younger can serve as an enlightening comparison of "then and now." Scenarios can be used to elicit trainees' thoughts and show how they might respond to a situation differently. For instance, how would they handle a customer complaint about slow service if they were the customer service manager? Answers would be influenced by the trainees' perceptions of customer demands based on their own expectations as customers. The Millennials are known for low tolerance of delays in customer service. They may take the complaint more seriously than the Xers. Regardless of their orientation, sharing the information opens trainees' minds to other perspectives and helps them in dealing with real customers that could be part of any generation.

Although matching training activities to generational learning preferences has advantages, ultimately it is the learning objective, not the learning preference, that matters most in determining the delivery modality. Employing a variety of modalities promote learning in all trainees by stretching them beyond their comfort zone (Johnson and Romanello, 2005, 216). At times you need to apply an instructional strategy that is not comfortable to all. Showing sensitivity and giving encouragement will help your trainees feel relaxed and willing to participate. For example, older trainees tend to be more reserved and afraid of being embarrassed. During group discussions and activities, avoid calling on them directly or asking them to role-play. If everyone must play a role, let them take on a less prominent role, such as one of the employees being evacuated versus the safety officer, unless they volunteer.

15.2.5.2 An Example

Pretend that your organization has sold a computerized risk management information system to a client to replace a manual system. You are the technical expert responsible for conducting a series of in-person training events for your client's staff in the use of the system. You have planned an introductory session to provide an overview and additional sessions that cover advanced features. Your client's risk management department has employees from different cohorts. All of them will be attending the introductory session. Here is one example of how you might develop your instructional strategies; many variations are conceivable.

The trainees need the user manual for the system. The Silents and possibly the Boomers may want the printed manual, whereas the Xers and the Millennials may prefer the digital format. Make both formats available no later than the time of the training event—the Millennials that are motivated may want to jump ahead to take a peek at their "new toy." They would also like to have the manual accessible on their mobile device so they can refer conveniently to the parts they need.

At the training event, use an icebreaker to enliven the atmosphere and so that you get to know the trainees. Since they are not strangers to one another, the Silents should feel comfortable to participate. Do not forget to introduce yourself and your areas of expertise; you can do so without appearing arrogant. For the main portion of the training, instead of showing the features of the system directly on computer monitors or a projection screen right away, introduce the terminology in a slide presentation, along with a handout because the Silents would want it. The lecture will ensure that the Silents are not confused because you are moving too fast. If you sense that they may need more time to digest the material, tactfully suggest that you will work with them during the break. The slide presentation can include video clips that demonstrate features. Adding animation will make the lecture entertaining for everyone and particularly maintain the attention of the Xers and Millennials. After the lecture, reinforce learning with a live demonstration and hands-on practice of using the actual system. Boomers like to be problem solvers. Stimulate their thinking by asking questions such as what the best solutions are in transitioning data from the manual to the computerized system. Allow plenty of time for questions; the Xers may have the most questions. Assure the Xers that the information will be accessible afterwards on the Internet or intranet, as the case may be. They will enjoy an exercise to independently research which data analysis and reporting functions of the system will be the most important in improving efficiency of the department's daily workflow. To engage the Millennials, prepare a competitive game in advance. Something like *Who Wants to be a Millionaire* or *Wheel of Fortune* will be good since money motivates the Millennials. The game can be used for final review and include questions that test knowledge of the terminology, which data input into the system are populated in different modules of the system, and how the modules relate to one another functionally.

15.3 SUMMARY

In practicing andragogy—the art and science of teaching adults (Merriam-Webster Unabridged, 2013), trainers should recognize how characteristics of trainees associated with their generation may affect the way they learn new materials or solve problems. Four generations working and learning together offer a great opportunity to use peers as resources to achieve superior outcome for all. For example, an employee from the Silent generation can be matched with an employee that is a Millennial in a mentoring program. The mentor-mentee relationship can be two-way, whereby the older employee shares his or her many years' work and life experience with the younger employee, and the younger employee helps the older employee in working with new technology. The benefits of this type of cooperative learning extend beyond a training program.

REFERENCES

Age Discrimination in Employment Act of 1967, 29 U.S.C. §§ 621–634 (2000).
American Management Association. 2007. "Leading the Four Generations at Work." http://www.amanet.org/training/articles/Leading-the-Four-Generations-at-Work.aspx.
Bureau of Labor Statistics. 2012. "Employment Projections—2010-20." News Release USDL-12-0160. Bureau of Labor Statistics, United States Department of Labor. http://www.bls.gov/news.release/pdf/ecopro.pdf.
Cavallini, E., A. Pagnin, and T. Vecchi. 2003. "Aging and everyday memory: The beneficial effect of memory training." *Archives of Gerontology and Geriatrics* 37 (3): 241–257. doi:10.1016/S0167-4943(03)00063-3.
Johnson, S. A., and M. L. Romanello. 2005. "Generational diversity: Teaching and learning approaches." *Nurse Educator* 30 (5): 212–216.
Laberge, J. C., and C. T. Scialfa. 2005. "Predictors of web navigation performance in a life span sample of adults." *Human Factors* 47 (2): 289–302.
McMichael, A. J. 1976. "Standardized mortality ratios and the 'healthy worker effect': Scratching beneath the surface." *Journal of Occupational Medicine* 18 (3): 165–168.
Merriam-Webster Unabridged, s.v. "Andragogy," accessed March 30, 2013, http://unabridged.merriam-webster.com/unabridged/andragogy.
Pew Research Center. 2012. "Older Adults and Internet Use." http://www.pewinternet.org/~/media//Files/Reports/2012/PIP_Older_adults_and_internet_use.pdf.
SeniorNet. 2013. "SeniorNet Fact Sheet." Accessed April 22. http://seniornet.org/index.php?option = com_content&task = view&id = 43&Itemid = 68.
Small, S. A., Y. Stern, M. Tang, and R. Mayeux. 1999. "Selective decline in memory function among healthy elderly." *Neurology* 52 (7): 1392–1396.

Thompson, C., and B. G. Sheckley. 1997. "Differences in classroom teaching preferences between traditional and adult BSN students." *Journal of Nursing Education* 36 (4): 163–170.

Wan, M. 2010. "Prolong the functional age: Tips for maximizing productivity, performance, and safety of the aging work force." *The Synergist*, March, 30–33.

Zemke, R., C. Raines, and B. Filipczak. 1999. "Generation gaps in the classroom." *Training*, November, 48–54.

16 Training a Multicultural Work Force

16.1 TRAINING IN A "FLAT" WORLD

In his book *The World Is Flat: A Brief History of the Twenty-First Century*, Thomas Friedman (2007) draws attention to the many effects of globalization. In this "flat" world, the importance of training a multicultural work force effectively is escalating as globalization leads to outsourcing and migration. Professionals in all fields have more opportunities to work with an employee population that is linguistically and culturally diverse. Multiple languages are heard and used in many work environments. What is the significance of a multicultural work force in the context of training? How can trainers work with this audience to achieve the desired results?

This chapter discusses a few cultural factors that may affect attitudes and behaviors and suggests best practices in training a multicultural work force to ensure understanding and compliance. Knowledge of cultural backgrounds that may influence trainees' learning preferences is helpful but it must not be used to put individuals in a "straitjacket." Milgram (2011), who carried out a series of experiments to measure conformity to common standards in France and Norway, points out the wide range of behavioral variations within one country.

16.2 UNDERSTANDING CULTURAL DIVERSITY

Culture is the values, beliefs, and practices shared by a group that may be identified in many ways such as country, institution, social class, or subgroups within these boundaries. Culture affects group members' daily living, their customary practices, and their interactions and means of communication.

In the United States, some of the industries with the highest percentage of workers whose native language is not English are construction, agriculture, and health care. These industries happen to be among the industries with the highest work-related injury rates. Effective safety training is critical to avoid illness, injury, or fatality but in these industries, accurately communicating information to workers that do not speak English fluently is a challenge. For example, even though health care professionals such as physicians are highly educated, they may lack conversational English skills (Hall et al., 2004, 122). This highlights the importance of applying culturally sensitive strategies in training to maximize learning.

Since every individual is different, it can be dangerous to base your interface with trainees on cultural stereotypes. Still, cultural norms shape individuals' responses to external stimuli and may directly affect trainees' attitudes and behaviors relating to training or a training topic. Divergence may emerge due to differences in cultures with regard to conformity, gender roles, uncertainty acceptance, and power distance.

16.2.1 CONFORMITY

Conformity is related to collectivism versus individualism (Bond and Smith, 1996, 124). Some cultures value conformity, while others nurture individuality. For example, the Milgram studies (2011) show that the French, compared with Norwegians, are more independent and resistant to group pressure. His experiments also illustrate that this cultural difference may exist not only between the West and the East but also among Western cultures. Trainees from a culture that highly values individuality may have a preference for self-directed and exploratory learning. Trainees accustomed to conformity may enjoy group discussions and activities. These are effective instructional strategies but the challenge for you as the learning facilitator is to prevent groupthink and promote critical thinking.

16.2.2 GENDER ROLES

A cultural norm that is related to conformity is gender (Eagly and Carli, 1981, 10). It is also an independent dimension of cultural values. Societies assign different roles to men and women. In male-dominant cultures, women tend to be more agreeable and dependent than men. The men are used to taking on decision-making roles. The converse is true in female-dominant cultures. These attitudes affect trainee-to-trainee and trainee-to-trainer interactions when the trainees are a mixed group or the trainer and a homogeneous trainee group are of the opposite gender. A trainee may not respect a trainer of the opposite gender if he or she believes that the other gender is inferior. If you are in such a situation, earn respect by building rapport and demonstrating your expertise in the subject matter.

16.2.3 UNCERTAINTY ACCEPTANCE

Some cultures are less tolerant of uncertainty than others. They try to minimize the uncertainty by setting strict rules and regulations. This approach may condition their members to be compliant but risk averse and lack the ability to handle unstructured situations that may arise. Members of uncertainty-accepting cultures may be better equipped to handle changes. In the training setting, trainees with a low uncertainty tolerance may be less willing to explore novel concepts and methods. Showing them how new applications have been proven effective and benefited others that use the applications would help these trainees feel comfortable about taking the risk of change. Note that risk behavior is predicted by risk perception and risk perception also varies across cultures (Lam, 2005, 186).

16.2.4 POWER DISTANCE

The influence of organizational factors may carry over from a trainee's country of origin. "Power distance is the extent to which the less powerful members of organizations ... accept and expect that power is distributed unequally" (Hofstede, 2013, under "Power Distance"). In high power distance cultures, organizations may exercise a high degree of supervision, whereas in lower power distance cultures, organizations may emphasize employee empowerment. Attitudes toward hierarchy, authority, and decision making may differ (Brett et al., 2006, 88). Employees that submit to much supervision can become too reliant on direction (Hsu et al., 2008, 32). This has two implications for training: how much direction must be given to a trainee during training and, more important, how much of the knowledge learned will be transferred to behavior on the job. If a trainee needs detailed direction and close supervision, learning may not move beyond the cognitive level to the behavioral level (33). This hurdle can be preempted by emphasizing application-level learning objectives and instructional strategies and ensuring that trainees understand why they must apply what they learn to do their job better.

16.3 AVOIDING CULTURAL PITFALLS

Best practices that provide guidance to avoid common pitfalls, motivate trainees, and enhance learning include the following:

- Respect the trainees.
- Apply cultural intelligence.
- Speak and write simply.
- Ensure proper translations.
- Use nonverbal techniques with discretion.
- Employ suitable instructional and communication strategies.
- Check for comprehension.

16.3.1 Respect the Trainees

Your instructional strategies are based in part on what you know about the background of the trainees such as age group, gender, education, and occupation. With a multicultural audience, it is necessary to pay attention to several subtleties.

Treat everybody equally. This may sound basic, but when some trainers detect a regional or "foreign" accent ("foreign" depending on where the training is conducted), they may break this rule inadvertently. Be careful not to ask questions such as "What is your nationality?" or react adversely to the person's accent or language style. When people are treated differently, they may feel that they are being discriminated against. Besides, nationality is not the same as ethnicity or country of origin. If you really need to know, wait for a suitable time during a conversation and tactfully ask something like "Where did you grow up?" Also, keep in mind that a second-generation immigrant may be influenced by the culture of their parents' country of origin and their degree of acculturation in their adopted country.

Prior experience of discrimination, socially or professionally, may affect a trainee's views toward training, which is another reason why actual and perceived equal treatment is important to help trainees build their self-confidence and enjoy the learning experience.

Immigrants who are working at entry level in their new country may have held supervisory positions in their home country. Regardless of their previous work experience, some may not be technologically savvy due to lack of access in their country of origin. Take care not to talk down to trainees because they hold entry-level positions or if they do not know how to operate certain equipment. Furthermore, the use of technology may be affected by field-dependent or field-independent cognitive learning styles, which in turn may be influenced by cultural and ethnic backgrounds (Wooldridge, 1994, 378).

With a heterogeneous group of trainees who speak different native languages, you should always use the language that has been chosen as the teaching medium. In other words, do not speak to a few individuals in a language that others present do not understand. If it is necessary to explain particular concepts to a number of trainees using their native language, arrange a time to do so outside the scheduled training event.

What if you have done your best in showing respect but someone among the trainees is prejudiced and hostile against another trainee? You will need to use the same tactics as you would handle a disruptive trainee. These are discussed in Chapter 17.

16.3.2 Apply Cultural Intelligence

Cultural intelligence reflects an outsider's ability "to interpret someone's unfamiliar and ambiguous gestures in just the way that person's compatriots and colleagues would, even to mirror them" (Earley and Mosakowski, 2004, 139). To adapt to varying cultural beliefs and practices that are unfamiliar, you must understand your own cultural biases and respect others' viewpoints. You are human and it is natural to have biases as long as you acknowledge them and not prejudice your actions.

When training a multicultural group, keep an open mind. Put yourself in another's shoes, which may require doing homework to discover what those shoes are like. For example, while it is common in the United States to address one's boss and colleagues by their first name, this practice may be considered disrespectful in another country. When you conduct training in another country, address your colleagues by their formal titles if that is customary in the host country.

One way to build rapport is to show respect by mirroring the customs and gestures of people from a different culture. This enhances the trainer-trainee relationship. For instance, you may notice that when a trainee gives you his or her business card, he or she holds the card with both hands and presents it to you as if it is an important, formal document. That is common in some Asian cultures. When handing out your business card in return, you can mimic the manner—do it smoothly!

Some customs and expressions are recognized and interpreted in the same way by multiple cultures. Explore common ground and use it to facilitate understanding. A warm smile rarely hurts.

16.3.3 Speak and Write Simply

In general use short sentences and simple words to express ideas. Keep in mind, though, that one-syllable words can be easily misheard and misinterpreted. Some training topics may require you to expand the basic vocabulary to include technical words. Be sure to explain those terms or provide a glossary. Trainees that speak a different native language may be fluent in your language in daily speech but not familiar with specialized terms.

Communication tools have been developed to help non-native English speakers converse effectively. One commercially available tool employs approximately 1,500 English words in simple, standard grammatical structure. The creator of this tool advocates against using humor with non-native English speakers; however, appropriate use of humorous stories in training helps retention. Be careful to avoid humor, metaphor, abbreviation, or anything that may confuse or insult the trainees. When unsure about whether certain words or expressions are offensive, it is better to avoid them. For example, debate goes on whether the term "flip chart" is offensive, so substitute it with "easel pad."

16.3.4 Ensure Proper Translations

In theory, providing materials in trainees' native language whenever possible ensures that they understand the content. Some government agencies and commercial vendors make warnings, posters, and training materials available in several languages. Translation services are also readily available. Software programs can generate translations as well. Yet a poor translation of materials can cause serious consequences or, at least, embarrassment.

In 2009, the United States Secretary of State presented a gift to the Russian Foreign Minister at their meeting in Geneva. The gift was supposed to be a mock "reset" button, symbolizing the mending of ties between the United States and Russia. The United States Department of State put on the button the word "peregruzka," which meant "overload" or "overcharge" instead of "reset," and made headlines in world news (BBC, 2009).

The problem with translations is that words have nuances and are prone to misinterpretation. Complex languages complicate matters. For instance, certain words in some languages such as English have more than one meaning, and different words may have the same pronunciation. Some terms in one language have no equivalent terms in another language.

Use the correct words in the right context and confirm cultural appropriateness. Translated materials should be complete, accurate, coherent, and grammatically correct. One way to verify accuracy and coherence is backward translation: After a document has been translated into a second language, it is translated back into the first language and compared with the original document to determine if the purpose and meaning are still intact.

Consider the literacy level of the trainees when translating written materials. In many countries, certain industries such as construction or agriculture have a large number of workers that are illiterate in their native language. In these instances, use more symbols and graphics instead of text and make sure that they are appropriate.

16.3.5 Use Nonverbal Techniques with Discretion

Although visual aids are helpful in overcoming language barriers, symbols and colors have different meanings in different cultures. As an example, Americans like to present a clock as a gift, perhaps as a symbol of structure or framework; to the Chinese, the phrase "giving a clock" phonetically means watching the recipient die!

Even symbols in international standards could be interpreted differently in different cultures (Smith-Jackson and Essuman-Johnson, 2002, 46). Some cultures are more attuned to using symbols than others. For instance, the Globally Harmonized System of Classification and Labeling of Chemicals (GHS) is an international standard for risk communication. The guiding principle for the development of the GHS is to have a system that is comprehensible worldwide through the use of pictograms. It has transpired, however, that implementing GHS in different countries may require cultural adjustments. Researchers have found that Japanese subjects were unable to recognize some of the GHS pictographic labels (Hara et al., 2007, 266). Consequently, when using a symbol in a visual, think about whether the symbol will be easily understood by the intended audience and will achieve the intended meaning. Symbols of tangible objects such as a building or a telephone are less subject to misinterpretation than symbols of abstract concepts. Instead of using only a pictogram, add text in the trainee's language.

Similarly, when using eye contact, facial expressions, body language, or other nonverbal techniques, ascertain that they are not objectionable to or interpreted differently by individuals with diverse cultural backgrounds. Several hand gestures common in the United States or the United Kingdom are considered insulting in some Latin American countries.

16.3.6 Employ Suitable Instructional and Communication Strategies

Training in some countries is based heavily on lecture. Trainees from such countries may be unfamiliar with other instructional strategies, such as role-playing or group discussion. Their cultural orientation toward reticence and formality and their concern with losing face may create barriers to learning. Explain the process to them in a positive and encouraging manner. Promote a supportive learning climate. Allow them plenty of time to adjust to the new learning experience.

When developing training media, take into account behavioral stereotypes, which are people's expectations of how things will behave in an environment (Chengalur et al., 2004, 295). The electrical wall switch is an example of an item whose construction follows behavioral stereotypes. In North America, the "up" position is "on" and the "down" position is "off." In the United Kingdom, the positions are reversed. These are the ways people in the respective countries would expect the switches to work. Based on this design principle, if your trainees are familiar with a language that reads from left to right, similar to English, you want to build the elements of a visual from left to right and place the most important information in the upper-left quadrant (Szul and Woodland, 1998, under "Basic Design Principles").

Modify communication style and pattern according to trainees' customs and traditions. For example, people tend to speak louder in some cultures than others. Preferences in direct versus indirect communication may differ (Brett et al., 2006, 87)—an important consideration when giving or soliciting feedback.

If the audience includes trainees who do not have a good command of the language used in training, speak slowly and convey one message at a time. Repeat as necessary. Avoid colloquial expressions and jargon. In a mixed audience where some are proficient in the language and others are not, schedule extra time later to work with those who may not fully comprehend the materials presented, so those who have a better understanding are not bored.

People who are not fluent speakers of a certain language may have an easier time communicating in writing in that language. They may also prefer face-to-face conversations over telephone conversations. Some words are easily misheard over the telephone. Speaking slowly and clearly is crucial.

16.3.7 Check for Comprehension

As always, encourage participation, solicit questions from trainees, and ask them questions. Some trainees may be culturally predisposed not to ask for help. Trainees who are not fluent in a language may be afraid of asking questions, of not being understood if they speak, or of making mistakes, so help them by building trust and encouraging feedback. For trainees who come from a culture where collectivism prevails, consider using a trained peer from the same culture as their role model.

Check for comprehension at regular intervals and verify that the response you receive is what you think it means. For example, nodding may convey different meanings in different cultures. Some trainees may nod or say "yes" to indicate that they are listening, not necessarily that they understand.

When selecting a training modality or deciding on the use of computer-based testing, consider whether the trainees' language ability or computer literacy might affect training effectiveness or test results. Provide an opportunity to perform a practice test if possible.

Deep-rooted traditions may inhibit some trainees from implementing what they have learned if it is drastically different from their culture. After the training, follow up with encouragement, support, and reinforcement.

16.4 SUMMARY

Training a multicultural work force is a challenge that requires you to overcome the barriers of linguistic diversity and cultural differences. It is also an opportunity to expand your horizons and help your employers or clients take advantage of the globalization of work. To achieve these purposes, you should enhance your cultural intelligence and apply culturally sensitive communication and facilitation skills.

ACKNOWLEDGMENT

This chapter is adapted from Wan (2011), a proceedings paper of the 100th Anniversary Professional Development Conference of the American Society of Safety Engineers. The information in the proceedings paper was originally published in the September 2010 issue of *The Synergist*, the magazine of the American Industrial Hygiene Association (reprinted with permission).

REFERENCES

BBC. 2009. "Button Gaffe Embarrasses Clinton." http://news.bbc.co.uk/2/hi/europe/7930047.stm.

Bond, R, and P. B. Smith. 1996. "Culture and conformity: A meta-analysis of studies using Asch's (1952b, 1956) line judgment task." *Psychological Bulletin* 119 (1): 111–137.

Brett, J., K. Behfar, and M. C. Kern. 2006. "Managing multicultural teams." *Harvard Business Review* 84 (11): 84–91.

Chengalur, S. N., S. H. Rodgers, and T. E. Bernard. 2004. *Kodak's Ergonomic Design for People at Work*. 2nd ed. Hoboken: John Wiley & Sons.

Eagly, A. H., and L. L. Carli. 1981. "Sex of researchers and sex-typed communications as determinants of sex differences in influenceability: A meta-analysis of social influence studies." *Psychological Bulletin* 90 (1): 1–20.

Earley, P. C., and E. Mosakowski. 2004. "Cultural intelligence." *Harvard Business Review* 82 (10): 139–146.

Friedman, T. L. 2007. *The World Is Flat: A Brief History of the Twenty-First Century*. Release 3.0. New York: Picador.

Hall, P., E. Keely, S. Dojeiji, A. Byszewski, and M. Marks. 2004. "Communication skills, cultural challenges and individual support: Challenges of international medical graduates in a Canadian healthcare environment." *Medical Teacher* 26 (2): 120–125.

Hara, K., M. Mori, T. Ishitake, H. Kitajima, K. Sakai, K. Nakaaki, and H. Jonai. 2007. "Results of recognition tests on Japanese subjects of the labels presently used in Japan and the UN-GHS labels." *Journal of Occupational Health* 49 (4): 260–267.

Hofstede, G. 2013. "Dimensions of National Cultures." Accessed April 27. http://www.geerthofstede.nl/dimensions-of-national-cultures.

Hsu, S. H., C.-C. Lee, M.-C. Wu, and K. Takano. 2008. "A cross-cultural study of organizational factors on safety: Japanese vs. Taiwanese oil refinery plants." *Accident Analysis and Prevention* 40 (1): 24–34. doi:10.1016/j.aap.2007.03.020.

Lam, L. T. 2005. "Parental risk perceptions of childhood pedestrian road safety: A cross cultural comparison." *Journal of Safety Research* 36 (2): 181–187.

Milgram, S. 2011. "Nationality and conformity." *Scientific American*. https://www.scientificamerican.com/article.cfm?id = milgram-nationality-conformity.

Smith-Jackson, T. L., and A. Essuman-Johnson. 2002. "Cultural ergonomics in Ghana, West Africa: A descriptive survey of industry and trade workers' interpretations of safety symbols." *International Journal of Occupational Safety and Ergonomics* 8 (1): 37–50.

Szul, L. F., and D. E. Woodland. 1998. "Does the right software a great designer make?" *T H E Journal* 25 (7): 48–49.

Wan, M. 2011. "Safety, seguridad, sécurité: Training a multicultural workforce." In *Safety 2011, Proceedings of the 2011 ASSE Professional Development Conference*. Des Plaines: American Society of Safety Engineers. CD-ROM.

Wooldridge, B. 1994. "Changing demographics of the workforce: Implications for the use of technology as a productivity improvement strategy." *Public Productivity & Management Review* 17 (4): 371–386.

17 Transitioning from Presenter to Facilitator

17.1 FOCUS ON TRAINEE ACHIEVEMENT

At the annual conference of a professional society, a guest presenter gave an outstanding talk. This scientist spoke for an hour on a hot topic in that profession. Unlike most other presenters at this conference, he was a dynamic speaker. He grabbed the audience's attention right from the beginning by telling a funny personal story of his blunders in his early career. He used video clips to show exciting new discoveries that were happening at his laboratory. Everyone in attendance was impressed! In the audience was the president of a local chapter of the society. She figured that members of her chapter would be interested to learn more about this topic and its application in their professional practice, so she invited the scientist to deliver a training course at the chapter's next meeting. At that meeting six months later, the scientist presented a four-hour training course. He was as engaging as before, but the impression was that he gave four presentations during the four hours. The materials were disorganized and disjointed. His jokes and video clips were entertaining; however, there was little audience interaction as he spoke almost nonstop. At the end of the day, attendees felt that they had watched a good performance but learned little.

As subject-matter experts, many incidental trainers have experience presenting at corporate meetings or professional conferences. Some are excellent presenters. They must recognize the distinction between a presenter and a trainer. A trainer's job is to facilitate learning. To that end, a trainer needs to understand how adults learn and apply that knowledge to help trainees along their way in the learning process. Adults are motivated to learn when the training course is related to their work and experience, when practical knowledge and skills are taught, and when they are involved in the planning process.

A presenter must consider the audience in order to successfully inspire, inform, persuade, or entertain its members. A facilitator goes beyond that and provides a learning environment focused on the performance objectives and how trainees can achieve those objectives. To be successful in this arena, a facilitator should have good communication and facilitation skills in addition to presentation skills. A presenter is not concerned about instructional design; a facilitator is, among other things. If you are like most incidental trainers, you probably do not have the support of an instructional designer—you are the instructional designer! As such, you will find it helpful to know a few basics of learning theories and instructional design.

17.2 UNDERSTAND LEARNING THEORIES

Several learning theories and their variations describe learning based on behavior, cognition, and environment. Cognition and environment are most relevant to the training setting. The fundamentals of these theories are presented in this chapter. The next chapter discusses practical applications in developing course materials and training activities.

17.2.1 INFORMATION PROCESSING AND COGNITIVE LOAD THEORY (CLT)

Cognitive psychology and neuroscience propose that a system of working memory holds and manipulates information for further processing later. When dealing with familiar material, the capacity of the working memory is not an issue because prior knowledge is stored in schemas in long-term memory. When new material is learned during training, the demand for information processing can be more or less than the capacity of the working memory. CLT is concerned with using optimal instructional control of the cognitive load to achieve the desired learning outcome effectively and efficiently (Paas et al., 2004, 1–2). Excessive overload or underload degrades performance. Consequently, when a learning situation has low information processing demands, the strategy would be to provide practice conditions that challenge the trainee. For a learning situation that has high information processing demands, it would be beneficial to reduce the load. As a result, training simple skills will require an approach different from training complex skills (Wulf and Shea, 2002, 207).

Since schemas become automated with practice and relieve the working memory from the processing burden, instructional design should encourage the construction and automation of schemas through practice (Paas et al., 2004, 2). Shea et al. (2000, 33) find that physical practice is more effective than observational practice but even observational practice alone is better than no practice. Alternating the two methods of practice is especially effective, possibly because of the social interaction that occurs between trainees and because a trainee can focus on one dimension of the task during observation and another dimension during practice (34). The method is also efficient because trainees are less fatigued both cognitively and physically. The amount of time to train two participants to a higher level of effectiveness is the same as training one participant with physical practice only. Additionally, for training that may expose trainees to risk of injury, such as athletic training, the risk during practice is reduced.

CLT distinguishes three kinds of loads of working memory (DeLeeuw and Mayer, 2008, 234; Gerjets et al., 2004, 39; Paas et al., 2004, 2):

- *Intrinsic:* Imposed by the number of information elements and their interactivity and is inherent in the material presented.
- *Extraneous:* Ineffective external load imposed by the way information and learning activities are presented that do not facilitate schema construction and automation.
- *Germane:* Effective external load that facilitates schema construction and automation and, therefore, contributes to learning.

Each of these loads does not represent the overall load and are not highly correlated (DeLeeuw and Mayer, 2008, 233–234). Intrinsic load can be measured by the amount of effort during learning, extraneous load changes response time, and germane load modifies difficulty rating. The three loads added together cannot exceed available memory resources. If course materials have high intrinsic load, the other two types of loads must be reduced for the trainees. High intrinsic load may be involved, for instance, in learning to operate complex machinery.

A heavy working memory load is especially detrimental for novices (Kirschner et al., 2006, 80). Instructional approaches that have been used to reduce intrinsic load include the following (Gerjets et al., 2004, 41–42; van Merriënboer et al., 2003, 6–7):

- *Part–whole sequencing:* Break down a complex task into subtasks and teach the subtasks; when trainees have mastered the subtasks, teach the complete task.
- *Simplified whole task:* Use a simplified version of the complete task and gradually increase the complexity.
- *Modular presentation:* Start with the total complex task but develop a modular solution procedure to limit the information elements to be processed concurrently.

The type of load in a given circumstance varies with the individual. What is extraneous for a novice may be germane for an experienced person. Within a manageable load, the goal is not to reduce the total load but to redirect the extraneous load to a germane load. In other words, aim at the lowest possible extraneous load and the highest possible germane load by considering the demands of the course materials and the presentation mode (Brünken et al., 2004, 131; Tuovinen and Paas, 2004, 150). The extraneous load is limited by using a pointer or other indicator to show trainees which part of the visual you are referring to; this avoids the need for an extensive visual search (Kalyuga et al., 2003, 26). By supplying trainees with worked-out examples or process worksheets before solving problems, time and mental effort during the training would be reduced, provided that the worked-out examples are structured at a cognitive load suitable for the trainees' experience level (Kalyuga et al., 2003, 26–27; Kirschner et al., 2006, 80; Paas and Van Merriënboer, 1994, 130). Information is more manageable in small chunks presented sequentially with breaks in between, rather than in a large segment (Kalyuga et al., 2004, 579).

Advocates of CLT claim that unguided or minimally guided instruction, as is often applied with the constructivist approach, is ineffective due to the high cognitive load, and may cause trainees to acquire incomplete or wrong knowledge (Kirschner et al., 2006, 84).

17.2.2 Constructivism

Learning theories focusing on the environment reflect the constructivist view that knowledge is constructed when new information comes into contact with existing knowledge. Constructivism emphasizes knowledge construction, not reproduction. Learning is an active process whereby "learners are active sense makers who seek to build coherent and organized knowledge" (Mayer, 2004, 14). The responsibility of the trainer is to encourage creativity and self-direction that will enable the trainee to build on existing knowledge, which is the result of cultural background and prior experience. The trainee plays an active role in a process of "learning by doing." As the trainee succeeds in completing simple tasks, he or she would gain confidence to tackle complex tasks. Instructional design should support such a learning process.

According to Bauersfeld (1992, 472), in a constructivist approach the trainee would develop competence through experimenting and interacting with the trainer and other trainees. Interaction is influenced by the situation and the individual's background. Bauersfeld contrasts an "object-centered or product-oriented" strategy with a "construction-oriented" approach (479–480). The former is concerned with comparison of trainee performance with an "objective truth" that is set as the standard and does not consider the trainee's construction of meaning. The latter is focused on the process of constructing and ascribing meaning more than the correct solution of the problem.

Constructivism emphasizes activities and collaboration through which trainees with diverse backgrounds learn through sharing experiences and viewpoints. It should not mean that all instructional methods used must include only hands-on or discussion activities (Mayer, 2004, 15). Mayer views active learning as comprising cognitive and behavioral activities. A pure discovery approach may lead to failure of a trainee to select information relevant to the subject matter, which is a cognitive activity (17). He suggests that the most genuine approach to constructivist learning is learning by thinking and guided discovery may be more appropriate for constructivist learning than pure discovery. Brown et al. (1989) propose a theory of situated cognition. Activity and perception at a nonconceptual level precedes conceptualization. Learning is a process of enculturation supported partly through social interaction, discussion, and reflectivity (40). In a study of skilled workers, Billet (1993, 21) finds that workers play a highly interactive role in the learning process through observation, consultation, and collaboration. Slower learning demonstrated by workers that are isolated illustrates the importance of guided learning and situated learning (23). Smith (2003, 81) suggests that computer-mediated communication can provide a social environment for isolated workers.

Conceptual growth occurs when a perspective of solving a problem is challenged and a different perspective is presented and considered (Greeno and van de Sande, 2007, 15). Through a process of "constructing perspectival understanding" (17), interaction occurs among members of the group, including the trainer, and results in a new perspective being accepted; therefore, the trainer may learn from the trainees as well. Since the new perspective may not be the original perspective intended by the trainer, he or she must rethink the problem and use constructive listening while appraising different perspectives (19). This highlights the importance of listening skills for an effective trainer and facilitator.

An implication of the constructivist approach is that traditional testing methods may not be appropriate for assessment of thinking, understanding, and learning (Prawat and Floden, 1994, 47). For instance, evaluation of learning may take the form of assessing both the trainees' problem-solving process and the solutions they propose (Jonassen, 1997, 86).

17.3 APPLY INSTRUCTIONAL DESIGN

The learning theories provide direction for efficient and effective instructional design. As a trainer and facilitator, you want to balance guidance and flexibility to achieve the optimal learning outcome. Options in instructional design can be evaluated by how much they promote cognitive processing (Mayer, 2004, 17). The effectiveness of a particular design can vary with the experience of the trainee. Experience changes the schemas that can be brought from long-term memory to integrate with new information in the working memory and, therefore, the cognitive load (Kalyuga et al., 2003). For a novice, guided methods of instruction such as lectures, tutorials, and worked-out examples compensate for the lack of prior knowledge and schemas relevant to a task or situation. They help the novice connect interacting elements to form schemas (24, 29). For a trainee with experience in the subject matter, the same instructional guidance hinders learning by loading the working memory unnecessarily. Instructional design that promotes "learning by doing," such as practice, role-playing, problem solving, and discovery, encourages deeper understanding and is suitable for trainees with prior knowledge, whether they have acquired the knowledge before the training or in the earlier part of the training.

Ultimately training must support organizational goals, which means that it must be connected to real-world problem solving. In today's workplace, problems are ill-structured (Gott, 1988, 162). As technologies change, manual tasks decrease and mental tasks increase. Critical elements of task performance in mental tasks are less observable and not articulated. Instruction should be designed to enable trainees to progress through "*situated, supported*, and carefully devised and *sequenced* learning experiences" (163). Situated means that knowledge and skills are taught in the context of solving actual problems. Support might include articulation of decision-making processes or manipulation of devices, provided to the trainee throughout the learning process and sequenced to match the stages of progression toward mastery of the subject matter.

17.4 HONE COMMUNICATION SKILLS

Much learning takes place through interactions. A good trainer and facilitator should have effective communication skills.

17.4.1 TWO-WAY COMMUNICATION

Communication is the creation or exchange of understanding between sender and receiver (Longest and Darr, 2008, 571–575). Ideas of the sender are encoded in words and symbols and sent through channels of communication to the receiver. The receiver's life and work experience and frame of reference affect how the message is decoded. Frame of reference includes culture, biases, beliefs, and values. Effective communication is established when a decoded message means the same as the one intended by the sender.

A message may not have the same meaning to the receiver as to the sender unless the sender anticipates and overcomes possible barriers to communication.

Some of the communication barriers that may exist between trainer and trainee are culture, education, socioeconomic background, or simply the way some trainees interact with others. Occasionally organizational factors such as human resource issues may play a part. Intonation, or the lack thereof in written communication, may cause misinterpretation. The same sentence uttered with stress placed on different words may have different meanings. Barriers result in misinterpretation or selective listening. For the training to be productive, you must gauge from trainee reactions whether your message is received as intended or needs repetition in a different way to be understood. A trainee's reaction may be verbal or nonverbal, such as a nod or a smile. Take care when decoding a message to consider the cultural background that is behind what a trainee says or does. Other ways to overcome communication barriers include devoting time and attention to the message, tailoring words to the receiver, encouraging free flow of communication, repeating a message using multiple communication channels, and practicing active listening skills.

17.4.2 Active Listening

Hearing is not listening. A sound heard may have no meaning. Listening requires thoughtful attention. The Chinese character for "listen" includes both of the characters for "ear" and "heart."

Always listen carefully to be sure that you fully understand and have not missed key points. To show that you are paying attention to a trainee while he or she speaks, make eye contact and lean forward slightly. Indicate that you are interested in and focused on what he or she is saying. Keep an open mind and evaluate new ideas objectively. Respond with appropriate facial expressions, questions, and summary. If another trainee tries to interrupt, politely request that person to hold off momentarily until the first trainee has finished speaking. Check back with the second trainee if what he or she wanted to say is related to the points under discussion. If it is, ask the person to express his or her thoughts. If not, let the individual know that you would get back to him or her when the current discussion is over—be sure to keep your promise.

Make active listening a habit by remembering and practicing the acronym LISTEN:

Lean slightly toward the speaker.
Ignore distractions.
Summarize to ensure understanding.
Toss out hasty judgments.
Engage the speaker with eye contact and show empathy.
Notice nonverbal nuances—the speaker's and yours.

17.5 PRACTICE QUESTIONING TECHNIQUES

Chapter 9 discusses several techniques in asking and responding to questions.

Asking questions is useful for several purposes. It grabs attention. It stimulates thinking. It draws in those who are shy. It checks for understanding of the course materials. To achieve these purposes, ask open-ended questions with "who,"

Transitioning from Presenter to Facilitator

"what," "when," "where," "why," and "how." Cover one issue in one question. If the individual or group asked seem to have difficulty finding an answer, provide hints. Listen actively to the responses. Repeat a response if you are not sure that everyone can hear it. Request the respondent to clarify if the response is not clearly stated. Comment on the response or ask a follow-up question if the response is incomplete or brings up a new point.

What is a good response to a question that you are asked? Your answer should address the question. As mentioned in Chapter 9, you may ask a return question that is linked to the original question. Use examples and illustrations to clarify your point, especially when it appears that a trainee has posed the question because he or she does not understand something you have tried to convey. Do not make the questioner feel stupid. Use tone, eye contact, and facial expressions to show encouragement. When there is not enough time to entertain all questions, rotate your selection of questioners to allow trainees in different sections of the room to speak. Inform the other questioners that you can speak with them during the break or at the end of the training event.

17.6 HANDLE DISRUPTIVE TRAINEES

Sometimes you may have disruptive trainees. They usually do not interfere with the training deliberately. They are just acting the way they normally do; therefore, do not take it personally. However, since they may negatively affect the learning environment, you need to deal with the situation when it becomes a problem. A general intervention strategy is to describe your observation and ask the disruptive trainee to discontinue what he or she is doing without putting down the person. Be assertive and tactful.

Disruptive behaviors are manifested in different ways. Some trainees do not say anything or appear reluctant to participate in activities. There are several reasons why they are like that. A person may be shy or introverted by nature. It is also possible that cultural background or language ability creates a barrier. You can draw this trainee in with a direct question. Address the person by name and ask a relevant question, for example, "How do you think the procurement department can save money?" Make the question relatively easy to answer so you do not intimidate him or her. Show encouragement and support to help the trainee express his or her opinions. Allow enough time for thinking. Another type of trainee who does not participate is the resistant learner (Sims and Sims, 2006, 284). His or her behavior may be due to lack of confidence, especially in the use of new technology or equipment. The solution lies in helping them build confidence, perhaps by dividing materials into smaller chunks so that they can "eat the elephant one bite at a time." Still another type is the result of the ubiquity of mobile devices. Some trainees might concentrate on using their devices instead of listening or participating in training activities. This "addictive" behavior is getting out of control even in college classrooms (Hammer et al., 2010, 299). When you are working with adult trainees, imposing penalties is not the answer. Plausible solutions for trainers are to be more engaging and to integrate mobile technologies into instructional strategies (301).

The opposite are those that respond to every question and volunteer to perform every activity. Participation is good but too much of it by one trainee is detrimental to the group. You need to ask the person to let others have a chance to comment or participate. For instance, you might say, "You have many great ideas. We appreciate your contribution. We need to hear from others as well. Would you please hold off your comments until the rest of the group is heard?"

Trainees that are know-it-alls are disruptive because they are not open to others' ideas. Do not argue with them. Instead, thank them for sharing their opinions and move on (Toastmasters International, 2008, 41).

You may also need to "declaw" crabs (Toastmasters International, 2008, 41). They are eternal complainers. If they have a valid complaint and there is a feasible solution, fix the problem. If not, explain to them why and continue with the training.

Another situation that calls for intervention is when one trainee makes hostile remarks about another or repeatedly puts down the ideas of another. The cause could be organizational or cultural issues but regardless, you must step in to preserve a positive learning environment. You might say to the individual, "I ask that you explore X's position with an open mind before dismissing it. That way X will be heard."

17.7 GIVE HELPFUL FEEDBACK

When you must handle a disruptive trainee, you are in effect giving one type of feedback that, hopefully, is not needed often. Giving feedback to trainees, however, is an important duty of a trainer and facilitator and should be done throughout a training course. This applies to in-person training and virtual training, trainer-led or self-directed.

If given tactfully, both positive and negative feedback promotes learning. Give praise when it is due, for example, when trainees solve a complex problem brilliantly. Negative feedback is needed to correct errors and should be given in a constructive and nonjudgmental manner. Focus on one or two major points at a time. Be specific about what is wrong and explain the reasoning clearly. Whenever practical, provide the feedback as soon as the error has occurred.

Feedback can be nonverbal. Your voice and body language send subtle messages to trainees. If you sound or appear irritated when responding to a trainee's question, you may give an impression that the question is inappropriate or stupid even though that is not your intention. Avoid any chance of creating misperceptions.

17.8 SUMMARY

The "job" of a trainer and facilitator is more than presenting a lecture or even organizing activities. It requires knowledge and skills that a presenter may not possess and does not need, such as instructional design or questioning techniques. With practice, you will be able to apply these skills smoothly and naturally. Your goal is to produce instructional materials that are educational and entertaining and to deliver training in an enthusiastic and engaging manner—in other words, to achieve the four E's of training. That is the discussion in the next chapter.

REFERENCES

Bauersfeld, H. 1992. "Classroom cultures from a social constructivist's perspective." *Educational Studies in Mathematics* 23 (5): 467–481. doi:10.2307/3482848.

Billett, S. 1993. "Authenticity and a culture of practice within modes of skill development." *Australian and New Zealand Journal of Vocational Education Research* 1 (2): 1–28.

Brown, J. S., A. Collins, and P. Duguid. 1989. "Situated cognition and the culture of learning." *Educational Researcher* 18 (1): 32–42. doi:10.2307/1176008.

Brünken, R., J. L. Plass, and D. Leutner. 2004. "Assessment of cognitive load in multimedia learning with dual-task methodology: Auditory load and modality effects." *Instructional Science* 32 (1/2): 115–132. doi:10.1023/B:TRUC.0000021812.96911.

DeLeeuw, K. E., and R. E. Mayer. 2008. "A comparison of three measures of cognitive load: Evidence for separable measures of intrinsic, extraneous, and germane load." *Journal of Educational Psychology* 100 (1): 223–234. doi:10.1037/0022-0663.100.1.223.

Gerjets, P., K. Scheiter, and R. Catrambone. 2004. "Designing instructional examples to reduce intrinsic cognitive load: Molar versus modular presentation of solution procedures." *Instructional Science* 32 (1/2): 33–58. doi:10.1023/B:TRUC.0000021809.10236.71.

Gott, S. P. 1988. "Apprenticeship instruction for real-world tasks: The coordination of procedures, mental models, and strategies." *Review of Research in Education* 15: 97–169. doi:10.2307/1167362.

Greeno, J. G., and C. van de Sande. 2007. "Perspectival understanding of conceptions and conceptual growth in interaction." *Educational Psychologist* 42 (1): 9–23. doi:10.1080/00461520701190462.

Hammer, R., M. Ronen, A. Sharon, T. Lankry, Y. Huberman, and V. Zamtsov. 2010. "Mobile culture in college lectures: Instructors' and students' perspectives." *Interdisciplinary Journal of E-Learning and Learning Objects* 6: 293–304.

Jonassen, D. H. 1997. "Instructional design models for well-structured and ill-structured problem-solving learning outcomes." *Educational Technology Research and Development* 45 (1): 65–94. doi:10.1007/bf02299613.

Kalyuga, S., P. Ayres, P. Chandler, and J. Sweller. 2003. "The expertise reversal effect." *Educational Psychologist* 38 (1): 23–31.

Kalyuga, S., P. Chandler, and J. Sweller. 2004. "When redundant on-screen text in multimedia technical instruction can interfere with learning." *Human Factors* 46 (3): 567–581. doi:10.1518/hfes.46.3.567.50405.

Kirschner, P. A., J. Sweller, and R. E. Clark. 2006. "Why minimal guidance during instruction does not work: An analysis of the failure of constructivist, discovery, problem-based, experiential, and inquiry-based teaching." *Educational Psychologist* 41 (2): 75–86. doi:10.1207/s15326985ep4102_1.

Longest, B. B., Jr., and K. Darr. 2008. *Managing Health Services Organizations and Systems.* 5th ed. Baltimore: Health Professions Press.

Mayer, R. E. 2004. "Should there be a three-strikes rule against pure discovery learning? The case for guided methods of instruction." *American Psychologist* 59 (1): 14–19. doi:10.1037/0003-066X.59.1.14.

Paas, F., A. Renkl, and J. Sweller. 2004. "Cognitive load theory: Instructional implications of the interaction between information structures and cognitive architecture." *Instructional Science* 32 (1/2): 1–8.

Paas, F., and J. J. G. Van Merriënboer. 1994. "Variability of worked examples and transfer of geometrical problem-solving skills: A cognitive-load approach." *Journal of Educational Psychology* 86 (1): 122–133.

Prawat, R. S., and R. E. Floden. 1994. "Philosophical perspectives on constructivist views of learning." *Educational Psychologist* 29 (1): 37–48.

Shea, C. H., D. L. Wright, G. Wulf, and C. Whitacre. 2000. "Physical and observational practice afford unique learning opportunities." *Journal of Motor Behavior* 32 (1): 27–36.

Sims, R. R., and S. J. Sims. 2006. "Maximizing learning outcomes in training and development: The critical role of learning and learning styles." In *Learning Styles and Learning: A Key to Meeting the Accountability Demands in Education*, edited by R. R. Sims and S. J. Sims. New York: Nova Science Publishers, pp. 259–291.

Smith, P. J. 2003. "Workplace learning and flexible delivery." *Review of Educational Research* 73 (1): 53–88. doi:10.2307/3516043.

Toastmasters International. 2008. *From Speaker to Trainer: Coordinator's Guide*. Rancho Santa Margarita: Toastmasters International.

Tuovinen, J. E., and F. Paas. 2004. "Exploring multidimensional approaches to the efficiency of instructional conditions." *Instructional Science* 32 (1/2): 133–152. doi:10.1023/B:TRUC.0000021813.24669.62.

van Merriënboer, J. J. G., P. A. Kirschner, and L. Kester. 2003. "Taking the load off a learner's mind: Instructional design for complex learning." *Educational Psychologist* 38 (1): 5–13.

Wulf, G., and C. H. Shea. 2002. "Principles derived from the study of simple skills do not generalize to complex skill learning." *Psychonomic Bulletin & Review* 9 (2): 185–211. doi:10.3758/bf03196276.

18 Achieving the Four E's of Training

18.1 THE FOUR E'S OF TRAINING

The keys to successful training outcome are the four E's of training: educational and entertaining materials plus enthusiastic and engaging trainer. Trainees remember better when the materials are presented in a manner that is educational and entertaining, and the trainer is enthusiastic and engaging.

18.2 EDUCATIONAL AND ENTERTAINING MATERIALS

Three items you should prepare besides the course materials are agenda, tent cards, and your self-introduction. The agenda helps you sequence topics in a logical flow and manage time. Schedule breaks about every hour. Some of them can be 5-minute stretch breaks. If you need to change the room setup for different segments of the event, make sure you allow enough time on the agenda. Whereas some trainers put the time for each segment on the agenda that is distributed to trainees, others put the time on their own copy only along with detailed process notes. The process notes describe tools and techniques that will be used for each topic. Use a font size of at least 14-point for all your notes so you can refer to them discreetly. This includes any notes for a slide presentation. The agenda is a guide. If it turns out that a certain topic generates much trainee interest or needs further elaboration, you can modify the agenda or you may even need to change the planned instructional strategy. For example, if you see that trainees are getting tired during a lecture, you can have them do a group activity. Tent cards with trainees' names help you and the trainees address people by name. You may wonder why you need to prepare your own introduction since you know who you are. Writing it out will make sure that you cover the main points, which are your background and experience that qualify you to conduct this training course. Type the introduction in sentence case in 14-point Times New Roman or larger with double-line spacing. If any unusual name is difficult to pronounce, include the pronunciation in the introduction. Read it out loud to see how it sounds and check the time. It should not last longer than 60 to 90 seconds.

The following sections discuss development of materials for specific activities.

18.2.1 ICEBREAKER

Icebreakers are an excellent way to start the training event. If the trainees are meeting one another for the first time, an icebreaker enables them to get acquainted. Even if they are coworkers, they may pick up tidbits or discover common ground that foster team spirit.

Identify the icebreaker you will use during a training event and assemble necessary materials. For any training activity, always prepare more materials than the number of trainees you anticipate. It is better having leftover than not having enough in case people show up at the last minute.

A large variety of icebreakers are available on the Internet or from other resources. In choosing an icebreaker, consider the size of the group and time. An icebreaker for a 1-hour program should not take more than 5 to 10 minutes (Toastmasters International, 2008, 11). Another consideration is whether trainees already know one another. Some icebreakers are more suitable for strangers than coworkers.

One activity that not only "breaks the ice" but also promotes commitment to the training objectives is to ask each trainee to share with one or more fellow trainees (a) why he or she is attending the training event, (b) what he or she expects to learn, and (c) what the benefits are, personally and professionally. Have each subgroup sum up three benefits and report back to the entire group. Summarize the benefits on an easel pad and post it in the room as a reminder throughout the training event.

18.2.2 MULTIMEDIA PRESENTATION

In this age of technology, it does not take much extra resource to present training materials in more than one modality. Does it mean the more, the better? That depends on what mental load the materials impose on the trainees. The cognitive load theory (CLT), discussed in Chapter 17, is significant in identifying best practices in the design of multimedia instruction.

The cognitive theory of multimedia learning (CTML) is an extension of the CLT. Effective working memory can be increased by presenting information in dual modalities, such as aural and visual (Moreno and Mayer, 2002, 163). For example, it is more effective to use a graphic with spoken text than written text because graphic and written text are both visual (Ginns, 2005, 326; Kalyuga et al., 2003, 26). Experiments on teaching the operation of a bicycle tire pump using narration before or concurrently with animation find that the concurrent presentation enhances learning outcome (Mayer and Anderson, 1991). It is also more effective than using narration only or animation only. The investigators conclude that animation should be accompanied by complementary narration. This dual-coding benefit is due to the split-attention effect when more than one source of information must be integrated to make the whole concept intelligible (Kalyuga et al., 2004, 578–579; Mayer and Moreno, 1998, 318). When two sources are not presented together, the working memory load is heavy because one piece of information must be held in working memory waiting for the other piece to be processed and integrated. Attention is split between the two sources. The remedy is to integrate the materials physically or temporally.

The redundant effect is related to the split-attention effect but operates differently. Redundant materials do not need to be mentally integrated. Each piece of information is intelligible by itself and describes the other in a different way. An example is presenting the same information orally and visually. Simultaneous presentation of redundant materials can have a detrimental effect on learning due to overloading the working memory because both sources of information are unnecessarily integrated (Kalyuga et al., 2004, 578–579; Mayer et al., 2001, 195). They should be presented sequentially so that the second presentation reinforces the first (Moreno and Mayer, 2002, 162). In some cases the redundancy can be eliminated. The common practice in multimedia instruction, especially in asynchronous e-learning, to provide identical spoken and written material simultaneously should be avoided (Kalyuga et al., 2004, 580). However, redundancy is dependent on the current level of knowledge and skill of the trainee. What is redundant for an experienced trainee may not be redundant for a novice (Kalyuga et al., 2000, 135). This is why the ability to customize information for trainees in virtual training, as discussed in Chapter 20, is beneficial.

Another concept is the coherence effect. Adding extraneous materials, even if they are interesting, distracts trainees and results in poorer performance due to priming of inappropriate prior knowledge from long-term memory as the organizing schema (Mayer et al., 2001, 196).

Multimedia instruction should foster information processing with the least amount of mental load. To summarize, these are the design implications (Ginns 2005, 326; Kalyuga et al., 2000, 2004; Mayer et al., 2001):

- Present simultaneously information that needs to be integrated for comprehension, for example, nonidentical oral and written materials.
- Deliver redundant materials in different modes sequentially; eliminate the redundancy if the pace of the presentation is fast and cannot be controlled by trainees.
- Do not add irrelevant material.
- For split-source graphic and text, present text in auditory form; provide a mechanism for experienced trainees to turn it off.
- Render material that has text only in aural or visual form but not both concurrently.
- Break up textual material into small segments separated by time breaks.

18.2.3 Self-Guided Discovery

The constructivism theory emphasizes interaction as central to the learning process. Workplace research also suggests that a typical worker prefers to learn within a socially constructed context and is not self-directed (Smith, 2003, 80). With the increasing popularity of distance learning, not to mention telecommuting, how is learning affected? While not all distance learning is self-guided, many courses are designed that way. Employing the instructional strategy of self-guided discovery does not mean that you leave the trainees all by themselves. You must help trainees adjust to this delivery mode and improve self-directedness. You would suggest learning strategies, provide resources, and give feedback. The trainee should be comfortable with his or her increased responsibility in the learning process to make it meaningful. Some may need more help than others.

According to Garrison (1997, under "A Comprehensive Model"), self-directed learning has three dimensions: self-management, self-monitoring, and motivation. Self-management includes identifying learning goals and managing resources that will help achieve those goals. Self-monitoring is the process whereby the trainee assimilates new and previous knowledge, assesses intermediate learning outcomes, and modifies strategies if necessary to achieve desired outcomes. Motivation refers to the motivation to take on the task of self-guided discovery and the motivation to complete the task to accomplish the learning goals. Motivation is high when trainees perceive that the learning goals are worthy and achievable and is influenced by such factors as self-efficacy and the perception of value versus effort.

Self-directedness means a trainee can identify and know how to pursue his or her learning goals independently as well as by reaching out to others in a community of practice. You want to provide a learning environment where interaction can take place between you and the trainees and among the trainees themselves. Means of communication are available through blogs, discussion forums, and group collaboration tools and you need to set these up. They are essential in virtual training. They are valuable for in-person training to encourage communication outside of the training event and continued learning. Some trainees may require guidance on how to make the most out of these features. You must provide directions that may include text or video tutorials.

18.2.4 Role-Playing

As explained in Chapter 5, role-playing gives trainees an opportunity to experience a situation first-hand. It is particularly appropriate for training in the affective domain. Your advance preparation includes planning the scenario and characters, deciding how you will present them during the training event, and putting together requisite materials. Will you assign roles or ask for volunteers? Since some trainees may feel uncomfortable with playing a role, it is better to ask for volunteers, but avoid having the same volunteers for multiple activities. Describe the situation and characters clearly to the trainees, especially those who will have roles. After the role-playing, lead a discussion on what the players have done well and what might be improved in relation to the training topic. If time permits, you may repeat the role-playing with different players to give more trainees the chance to participate or illustrate another approach to performing a task or solving a problem. Do not repeat too many times or interest will subside; therefore, role-playing should not be used for summative assessment in a large group.

18.2.5 Worked-Out Example

A good strategy for instructional design is to sequence the materials to enable a smooth transition as knowledge advances. Worked-out examples can be combined with problem solving. One method is fading, or using examples that do not show complete solutions and having trainees supply the missing parts (Renkl et al., 2004, 2002). For example, an inventory clerk in the shipping and receiving department must follow ten steps to record and store an incoming shipment correctly. When training inventory clerks, a trainer provides a worked-out example of how a shipment of books is handled but deliberately places question marks in place of several steps on the list. The trainees must think about what those steps are. For more experienced trainees, the level of difficulty can be increased by omitting the question marks. Trainees must determine the missing steps and where they fit into the sequence of events. Another way of integrating a worked-out example with problem solving is the example-problem pair where a worked-out example is given first, followed by trainees' problem solving in an analogous situation. For instance, the inventory clerk trainees are shown the procedures for handling a shipment of books and then asked to describe the handling of another type of product. The level of difficulty is varied by changing the structure of the examples and solutions. It is easier when the situations are more similar, such as handling books and cutlery as opposed to books and perishable food.

You must decide what worked-out examples are appropriate for a training topic, whether to include a problem-solving element, and the level of difficulty. Carefully review, or ask a peer to review, the examples you plan to use to ensure that they are relevant and realistic. Determine how you will present them, which would depend on the nature of the training topic and the examples.

18.2.6 Problem-Solving Activity

Problem-solving activities can be educational and entertaining. From the educational standpoint, these activities are versatile and can be designed based on the information processing or constructivist approaches to learning. From the entertaining standpoint, they can be presented in many forms that add energy and excitement to the training event.

Problems are defined by four elements: problem domain, type, problem-solving process, and solution (Jonassen, 1997, 66). These elements can be well-structured or ill-structured, corresponding to the information processing approach or constructivist approach, and anywhere in between on this continuum. A well-structured problem requires the application of a set of clearly defined principles and parameters to arrive at convergent solutions. An ill-structured problem has multiple solution paths and multiple solutions. One or more of the elements of the problem are unknown or uncertain. The attributes of ill-structured problems are akin to real-world problems in the workplace (Gott, 1988, 162).

Jonassen (1997, 73–78) recommends the following steps for implementing well-structured problem-solving instruction:

- Review concepts, rules, and principles.
- Present a conceptual or causal model of problem domain in a visual representation.
- Model problem-solving performance in worked examples, including a description of the thought process.
- Present practice problems.
- Support the search for solutions by providing hints, if needed, and feedback.
- Reflect on the problem state and solution to reinforce learning.

To implement ill-structured problem-solving instruction, Jonassen (1997, 83–86) suggests the following steps:

- Articulate the problem context or situation.
- Set problem constraints such as time and budget.
- Identify real-world cases that engage the knowledge and skills taught.

- Compile knowledge base with evidence that represents alternative opinions.
- Present the problem and related information to the trainees.
- Support argument construction by asking trainees reflective judgment questions.
- Assess the problem solutions in terms of their viability.

A special point to note is the argument construction. When the learning objective is at the problem-solving level, you want the trainees to be able to articulate the arguments that support their proposed solutions and to base their judgment on correct facts and reasonable assumptions. They should distinguish facts from assumptions and opinions. Time-permitting, you should encourage them to research the issues themselves instead of relying only on the knowledge base you provide.

You can use well-structured and ill-structured problems in sequence, starting with the well-structured. Think through the problem from all perspectives to make sure the principles and parameters are indeed well-defined and there is only one suitable solution. Peer review will be helpful. Since ill-structured problems do not have one solution, they can be used to construct a case study for discussion or for debate.

18.2.7 Debate

Select a controversial statement or ill-structured problem related to the training topic. For instance, in a continuing education course for health care professionals, a statement might be "We should hire more international medical graduates to mitigate the physician shortage." Many debate rules exist. Below is an example of the basic procedure:

- Introduce the topic and debate rules.
- Form two teams of three trainees each.
- Let the rest of the group be observers and judges.
- Toss a coin to determine which team is affirmative and which team is negative.
- Allow the teams 15–20 minutes to prepare their arguments.
- Set time limits for statements and rebuttals, giving each team the same time.
- After each presentation, give the opposition a shorter time to refute.

Choose one or more debate topics before the training event. If there are many trainees and time is adequate, you can form more than two teams to debate different topics. Compile a list of the main arguments for and against the issues to make sure that they are covered by the teams' presentations or during the debriefing. Set the time limits for the presentations and determine judging rules for awarding points to the teams. A nicety is to have a prize or award that will be presented to the winning team.

18.2.8 Game

Game activities such as puzzles are frequently used in training. Most puzzles present well-structured problems when the problem, rules, and goal are stated unambiguously.

They can be used for different levels of learning objectives. Competitive games add excitement and may increase motivation. Before the training event, choose a game based on the group size, available time, and level of difficulty of the game (understanding how it is played and playing it). Prepare the game by incorporating content from the training topic. Many software applications and templates on the Internet are free and materials for some traditional games like bingo are inexpensive.

18.3 ENTHUSIASTIC AND ENGAGING TRAINER

Enthusiasm is contagious. You must be enthusiastic for your trainees to be enthusiastic and accept what you want them to learn. Enthusiasm comes from within and is hard to fake if you are not truly excited about the subject matter or trainees' success. When you are passionate, your enthusiasm is displayed through your words, voice, and body language. Your trainees will feel your emotions and be influenced by them. On your part, do not allow one or two trainees' behaviors to dampen your enthusiasm. For example, when you encounter disruptive trainees, employ intervention strategies discussed in Chapter 17 and stay calm.

Although some individuals seem to be naturally engaging, the ability to engage trainees' attention can be acquired by practicing effective communication, presentation, and facilitation skills. The efficacies of certain techniques are well established. Some of them are discussed in Chapter 9. Human interest stories connected to the training topic catch attention. Humor has a positive impact on learning retention (Garner, 2006, 179). Appropriate humor gives distance learning a human touch and enhances engagement (Anderson, 2011, under "Discussion"). It must be emphasized that what is appropriate varies with cultural and individual characteristics. That is also true for the choice of words and symbols. Avoid language bias such as "Hey, guys!" unless all trainees are men. Use simple vocabulary. Vocal variety and dynamic body language should support what you are saying. Body movements and acting can create dramatic effects that are memorable. Meanwhile, be mindful of the cognitive load when trainees must process multiple sources of information you present. An example is when you give a slide presentation. While showing a slide, talk about the points you are illustrating, not something unrelated.

Choose illustrations and comparisons that relate to trainees' existing knowledge and experience to make a point. Encourage trainees to explore how they can apply new concepts to their job. If the experience levels of the trainees vary, ask the more experienced to share their ideas. If all trainees lack experience, each can be assigned to research a different part of a group project and share the findings with the group. Trainees enjoy cooperative and active learning (Thompson and Sheckley, 1997, 168). Sitzmann et al. (2008, 287) find that instructor style and human interaction have strong influence on trainee reactions. That is why enthusiasm and engagement are impactful.

Discussions may be the most common activity during your training event and are vital to engaging trainees. To ensure that a discussion is focused on the learning objectives and not a casual conversation, discussions should be well organized and structured, whether they are general discussions or breakout sessions.

18.3.1 General Discussion

You may want to post ground rules at the beginning of the event along the following lines:

- Focus on the topic of discussion at hand.
- Analyze issues from all perspectives within the scope of the training topic.
- Support your viewpoints with evidence.
- Give every person in the group a chance to speak.
- Ask for clarification when in doubt.
- Respect others' opinions even if you do not agree.
- Offer constructive feedback, not criticism.

In your process notes you should have a list of points that fall under each topic. While you moderate the discussions, check that the major points are covered. If not, ask questions of the trainees about the missing items. If the group's opinion on a controversial subject is lopsided, stimulate thinking by opposing the group's point of view. This situation occurs more often in training that aims at changing behaviors. Stay on track and on time. At the end of a discussion, summarize the main points and conclusions. Provide feedback on the process and outcome.

A special format for discussions is the round robin or round table. It can be used for the entire group or for subgroups in breakout sessions. It is a good way to ensure that every trainee contributes ideas to the discussion. Members of the group sit in a circle or around a table. Going clockwise or counterclockwise, each member gives his or her opinion on the topic within a set time limit. The round robin is repeated until the topic is fully explored or at the end of the time allocated for the discussion.

18.3.2 Breakout Session

Depending on the size of the group and materials to be covered, you can hold subgroup discussions and assign each subgroup the same topic or different topics. The latter is preferred because everyone will learn more. Breakout sessions can efficiently cover several topics or subtopics at one time. The smaller group size allows more opportunities for sharing ideas and encourages trainees who are shy to participate in the discussion. The same ground rules as for the general discussions should apply to each subgroup. Divide trainees into subgroups of three to five by seating area or, better, randomly. Instruct them to select a leader who will represent them and report back to the entire group later. The leader also has the responsibility of encouraging everyone to participate. An alternative is to have different persons serve as leader and recorder. The leader manages the discussion and the recorder takes notes and reports to the entire group.

Assign the topics to the subgroups. Let them know their specific objectives and what they should report. For example, you may want each subgroup to present only two or three of their best ideas to the entire group for two minutes. Inform them of the time for the breakout session. Give them a warning three to five minutes before time is up and remind them that each person should contribute. Enforce the time limit.

Reconvene the entire group to receive the subgroup reports. As in general discussions, make sure that main points are covered, summarize, and provide helpful feedback.

18.4 SUMMARY

Preparation takes time. Do not wait until the night before the training event to assemble the materials. It is safe to pilot test or peer review new content or new format of presentation before implementation. Making sure that you have the four E's of training will help you direct an energized training event every time.

REFERENCES

Anderson, D. G. 2011. "Taking the 'distance' out of distance education: A humorous approach to online learning." *Journal of Online Learning and Teaching* 7 (1). http://jolt.merlot.org/vol7no1/anderson_0311.htm.

Garner, R. L. 2006. "Humor in pedagogy: How ha-ha can lead to aha!" *College Teaching* 54 (1): 177–180.

Garrison, D. R. 1997. "Self-directed learning: Toward a comprehensive model." *Adult Education Quarterly* 48 (1): 18.

Gott, S. P. 1988. "Apprenticeship instruction for real-world tasks: The coordination of procedures, mental models, and strategies." *Review of Research in Education* 15: 97–169. doi:10.2307/1167362.

Ginns, P. 2005. "Meta-analysis of the modality effect." *Learning and Instruction* 15: 313–331. doi:10.1016/j.learninstruc.2005.07.001.

Jonassen, D. H. 1997. "Instructional design models for well-structured and ill-structured problem-solving learning outcomes." *Educational Technology Research and Development* 45 (1): 65–94. doi:10.1007/bf02299613.

Kalyuga, S., P. Ayres, P. Chandler, and J. Sweller. 2003. "The expertise reversal effect." *Educational Psychologist* 38 (1): 23–31.

Kalyuga, S., P. Chandler, and J. Sweller. 2000. "Incorporating learner experience into the design of multimedia instruction." *Journal of Educational Psychology* 92 (1): 126–136. doi:10.1037/0022-0663.92.1.126.

Kalyuga, S., P. Chandler, and J. Sweller. 2004. "When redundant on-screen text in multimedia technical instruction can interfere with learning." *Human Factors* 46 (3): 567–581. doi:10.1518/hfes.46.3.567.50405.

Mayer, R. E., and R. B. Anderson. 1991. "Animations need narrations: An experimental test of a dual-coding hypothesis." *Journal of Educational Psychology* 83 (4): 484–490. doi:10.1037/0022-0663.83.4.484.

Mayer, R. E., J. Heiser, and S. Lonn. 2001. "Cognitive constraints on multimedia learning: When presenting more material results in less understanding." *Journal of Educational Psychology* 93 (1): 187–198. doi:10.1037/0022-0663.93.1.187.

Mayer, R. E., and R. Moreno. 1998. "A split-attention effect in multimedia learning: Evidence for dual processing systems in working memory." *Journal of Educational Psychology* 90 (2): 312–320. doi:10.1037/0022-0663.90.2.312.

Moreno, R., and R. E. Mayer. 2002. "Verbal redundancy in multimedia learning: When reading helps listening." *Journal of Educational Psychology* 94 (1): 156–163. doi:10.1037/0022-0663.94.1.156.

Renkl, A., R. K. Atkinson, and C. S. Grosse. 2004. "How fading worked solution steps works—A cognitive load perspective." *Instructional Science* 32 (1/2): 59–82. doi:10.1023/B:TRUC.0000021815.74806.f6.

Renkl, A., R. K. Atkinson, U. H. Maier, and R. Staley. 2002. "From example study to problem solving: Smooth transitions help learning." *The Journal of Experimental Education* 70 (4): 293–315. doi:10.2307/20152687.

Sitzmann, T., W. J. Casper, K. G. Brown, K. Ely, and R. D. Zimmerman. 2008. "A review and meta-analysis of the nomological network of trainee reactions." *Journal of Applied Psychology* 93 (2): 280–295. doi:10.1037/0021-9010.93.2.280.

Smith, P. J. 2003. "Workplace learning and flexible delivery." *Review of Educational Research* 73 (1): 53–88. doi:10.2307/3516043.

Thompson, C., and B. G. Sheckley. 1997. "Differences in classroom teaching preferences between traditional and adult BSN students." *Journal of Nursing Education* 36 (4): 163–170.

Toastmasters International. 2008. *From Speaker to Trainer: Coordinator's Guide.* Rancho Santa Margarita: Toastmasters International.

19 Directing an Energized Training Event

19.1 GET READY

Armed with educational and entertaining materials and being an enthusiastic and engaging trainer, you are almost ready for the training event. This chapter adds details about preparation and delivery. Attention to details will make your course materials and training event memorable. It will reflect on your professionalism and credibility. Most of the following discussions apply to both in-person training and videoconferencing.

19.2 PROMOTE ATTENDANCE

You have put in a lot of work to design and develop the training course. You know from the needs assessment that trainees will benefit from attending. This is no guarantee that they will actually attend. Mandatory attendance imposed by supervisors and organizational policies may dictate employees' attendance in some training courses but voluntary participation enhances learning outcome. To promote attendance, begin to publicize the training event to your target audience at least 8 weeks before the scheduled date. Use as many channels of communication as possible with available resources. Emphasize the benefit of attendance, such as improved job skills that save time or increase earnings. Perceived job and career utility and supervisor support predict training motivation, so promoting the training event also to the supervisors of potential trainees is advantageous (Clark et al., 1993, under "Discussion"). Make registration or requests for information easy by providing a contact name and phone number and inserting the relevant links in electronic materials.

The most convenient method to distribute the training event announcement is through intranet or e-mail, although e-mail delivery to recipients outside an organization can be unreliable. E-mail advertising outside the organization must also comply with applicable laws. Flyers and posters, physical or electronic, can be placed on company bulletin boards and distributed in areas like the cafeteria. When potential trainees come from different organizations or communities, you can reach many prospects through public and college libraries, chambers of commerce, associations, newspapers, and the broadcast media, depending on the training topic, target audience, whether the event is sponsored by a nonprofit organization, and if registration is free. The announcement needs to be repeated because people may have interest but take no action until they are reminded several times.

19.3 HANDLE LOGISTICS

Needless to say, you must prepare all materials and the physical or virtual environment that are directly related to training delivery. If you have invited guest speakers, you may want to request each of them to give you a biography. Most of the time you will receive generic information that they use for all occasions. Edit it to emphasize the speaker's expertise and type it up as you did for your own introduction. The introduction will be easy to read and you can maintain eye contact with the trainees while giving the introduction.

Supplies, equipment, or service you need from facilities personnel or third-party vendors should be ordered well in advance and reconfirmed as the date of the training event approaches. Monitor the expected number of trainees and adjust quantities of orders if necessary. Verify that trainees who have registered for the event have received all the information they need, including log-on instruction in the case of videoconferencing. The instruction may be accompanied by suggestions to prevent noise and distractions during the videoconference, such as putting a "do not disturb" sign on the door.

Use a checklist to make sure that everything is available or has been arranged. The list may include personal items to bring if you must travel to the site of the training event. A pocket-size clock with large numbers is good for checking time at a glance instead of having to look at your watch or other device during a lecture or activity.

19.4 ARRIVE EARLY

You should allow plenty of time to make sure that the room is correctly set up and all equipment is functioning properly. This applies to in-person and virtual training. For in-person training, you want to know that instructions you gave to hotel staff or facilities services for room and equipment arrangements are followed. For virtual training, even if you have used the same technology and service provider many times before, sporadic glitches can arise. In both cases, the need to double-check cannot be overemphasized.

When setting up a computer to be used for slide projection, be sure to verify the computer settings so that irrelevant sounds and pop-up windows will not appear at inopportune times. For example, if you are connected to the Internet and you leave an e-mail program running, turn off the audio notification for new messages. Watch out for the automatic update utilities of application programs on your computer and turn them off too. After setting up the projection equipment, check to ensure a clear view from the audience seating. Also check that your voice can be heard in the back of the room or if you need a microphone. Wireless lapel microphones are preferable to wired microphones because you can move around the room easier.

Trainers like to bring candies to training events for general distribution to trainees or as rewards to those who actively participate. If you have a bag of chocolates, do not put it next to the cooling fan outlet of a portable data projector on a table. In addition, if you use a portable data projector, be sure to set the keystone adjustment and zoom ring properly so that the image on the screen is a rectangle, not a parallelogram or trapezium. You may need to change the distance between the screen and projector to get the best result.

Directing an Energized Training Event   175

When you have an assistant or moderator, request that the person arrive early as well and go over how responsibilities are shared and coordinated, such as advancing slides during a slide presentation or handling questions from trainees. The rehearsal is crucial to the success of your training event.

How early is early depends on the venue and complexity of the setup. At a physical location, plan to have everything ready about an hour before the scheduled event. That way you are all set and relaxed before the first trainee arrives. Sometimes trainees show up 30 minutes or more before the start time. For virtual training, you should have all hardware, software, and Internet connections tested at least 15 minutes before the scheduled event. Participants of videoconferences tend to be late logging on, but it is fair that you are prepared for those who are prepared.

19.5   GREET TRAINEES

Besides avoiding the embarrassment of rushing around trying to make last-minute preparation in the presence of early birds, another reason for allowing plenty of time is so that you can greet the trainees as they arrive. Again, this is true whether you are in a physical or virtual classroom. Chatting with trainees in an informal atmosphere is an opportunity to build rapport. Trainees will no longer feel that you are a total stranger when you start the program. You understand some of the trainees' backgrounds and reasons for attending the training event. You can use this information to adjust the course delivery to fit their needs.

19.6   START ON TIME

You probably have the experience, more than once, of hurrying to get to a meeting on time and then found that the meeting did not start until 5 or 10 minutes after the scheduled time. It happens quite often and demonstrates lack of professionalism and organization skills on the part of the meeting planner. You should respect trainees' time. Delay in starting the program is unfair to those who arrive on time and cannot be justified by the tardiness of the others. Furthermore, you may not be able to cover all materials if you lose time. Incidentally, after a break not everyone will return to their seats on time. Begin speaking and people will settle down.

Start by extending a warm welcome. Introduce yourself briefly. That will give you the credibility you deserve when you begin the program. If a moderator introduces you, you should have provided the moderator with your introduction and reviewed it with him or her earlier.

If the moderator has not already done so or if there is no moderator, go over quickly housekeeping and ground rules. Emphasize that all phones and other noise-making devices must be in a silent mode and do not forget to set your own!

Ground rules should be established as well for videoconferencing. Reminder about eliminating noise and distractions is even more important since the physical environment of some trainees is unknown and uncontrolled.

Inform the trainees when and how they will be asked to complete a course evaluation for this event and why the evaluation will help them and you in the future.

19.7 PRESENT THE PROGRAM

What you present during the main portion of the training event is what you want trainees to remember and use in their job. Your goal is to effectively implement instructional strategies, employ presentation and facilitation skills, and provide the best learning experience for the trainees. The discussions and guidelines that follow relate to several aspects of course delivery that will help you achieve this goal.

19.7.1 Events of Instruction

Based on Gagné and Briggs (1979, 152–171), the following are nine steps of instruction. They are generic and can be used for most training events.

1. Gain attention: Start with a dramatic opening and icebreaker.
2. Inform trainees of the learning objectives: Reinforce what is in it for them.
3. Stimulate recall of prior learning: Review prerequisites.
4. Present the stimulus: Deliver course materials.
5. Provide learning guidance: Use demonstration, model, etc.
6. Elicit performance: Ask trainees to demonstrate learning in role-playing, etc.
7. Provide feedback: Praise good performance; suggest improvements.
8. Assess performance: Administer a test and give more feedback.
9. Enhance retention and transfer: Show how skills are transferred to the job.

When you review the learning objectives, explain what testing and assessment methods will be used to determine satisfactory completion of the course. Summarize concepts and solutions and check for understanding before leaving a topic. When moving from one topic to the next, make appropriate lead-in statements to transition smoothly, for example, "Now that we have learned about the psychology of social media, let's look at how it impacts our marketing efforts." Always maintain an atmosphere of openness conducive to learning.

Whether you are using any visual aids, you do not want to stand in one spot throughout the time you give a presentation. Move around the room to be close to the trainees and speak to them directly. Make sure that everyone can still hear and see what you are doing as you move around.

19.7.2 Interactive Methods

Your instructional strategies may include a number of interactive methods such as role-playing and case study. The following general guidelines can be used to implement various interactive methods:

- Explain the activity and procedures, including allocated time.
- Introduce background information or scenarios.
- Direct trainees to perform the activity.
- Solicit trainee reactions as performers and observers.
- Offer feedback on performance and process.

Directing an Energized Training Event 177

- Ask trainees to generalize principles for learning transfer.
- Invite trainees to apply principles to their job.

Directions should be specific and clear to the trainees. Time management when interactive methods are used is challenging. Watch the time closely during these activities.

19.7.3 VISUAL AIDS

The first rule about using visual aids is this: Do not expect visual aids to save a poorly prepared presentation! The second rule is that whenever you use technology, expect Murphy to drop by. Anticipate technical and other problems and have a "Plan B." For instance, if you use a data projector, it behooves you to have a spare lamp. However, if for whatever reason the projector or computer quits, you should be able to deliver the materials without the slides.

The following are tips for using two of the most popular visual aids, slide presentation and easel pad.

19.7.3.1 Slide Presentation

Chapter 6 has recommendations for slide design. Here are suggestions on delivery of the presentation (Brier and Lebbin, 2009, 360; Toastmasters International, 2008, 36).

- Use a remote control or an assistant to advance the slides.
- Talk to the trainees, not the slides.
- Do not read from the slides.
- Maintain eye contact with the audience.
- Reveal or highlight the point that you are discussing.
- Blank the screen when you are not referring to the slides.

Consider incorporating into the presentation a variety of features that hold trainees' attention, such as polling by the use of an audience response system to solicit opinions on issues of interest.

19.7.3.2 Easel Pad

In contrast to the slide presentation, easel pads are not suitable for a large group. They are handy for small group discussions. These are some suggestions (Toastmasters International, 2008, 35–36):

- Use water-based markers in dark colors such as black or blue.
- Write text in uppercase letters or small caps at least 1¼ inches tall.
- Stand sideways when writing so an audience can see what you write.
- Stop talking when you are writing, or recruit a scribe.
- Put one topic on one page.
- Leave every other page blank; turn over to a blank page when finished with a point.
- Stick index tabs made out of masking tape on the sides of the pages for reference.

After the discussion, you can post the pages around the room as reminders of the points or distribute them among subgroups for discussions in breakout sessions.

19.7.4 GUEST SPEAKERS

Introduce the guest speakers to the trainees. Do it with enthusiasm! If you can, memorize the introductions and do not read from your notes. Listen actively during their presentations so that you can better moderate the question-and-answer session that follows. Prepare one or two "seed questions" to ask in case no trainee submits a question after the guest speaker's presentation. You can also use the questions to direct the discussion. Give the guest speakers timing indicators and request that they adhere to their allocated time.

19.8 END ON TIME

If you were an instructor teaching undergraduate students, you may find that ending your classes early will get you higher ratings in the course evaluations. As an incidental trainer, you hope to have trainees who are more mature and appreciate the value of the training you provide. That does not mean more is better. Just as you should start the training event on time, you should also end it on time so that you do not interfere with trainees' plans, whether they are returning to work or doing something else after the training. Be available for those who may want to speak with you or need further assistance on the topics presented.

Toward the end of the program, remind trainees to submit the course evaluation. Summarize the main points covered in the course. Restate the benefits to motivate trainees to apply the knowledge and skills learned to their work. Let them know that you are available if they have questions later, and encourage them to keep in touch as you are interested in how they do in their job. End with a positive and powerful close.

19.9 SUMMARY

Most of the discussions in this chapter are relevant to traditional in-person training and virtual training via videoconferencing. Virtual classrooms have almost all the capabilities of physical classrooms, including "face-to-face" discussions and collaboration. The growth in the use of new technologies in training, from course development to delivery, is at an unprecedented pace. As an incidental trainer, you want to take advantage of what the technologies can offer to enhance learning experience and outcome. Virtual training and mobile learning are the new paradigm.

REFERENCES

Brier, D. J., and V. K. Lebbin. 2009. "Perception and use of PowerPoint at library instruction conferences." *Reference & User Services Quarterly* 48 (4): 352–361.

Clark, C. S., G. H. Dobbins, and R. T. Ladd. 1993. "Exploratory field study of training motivation: Influence of involvement, credibility, and transfer climate." *Group & Organization Management* 18 (3): 292+. Academic OneFile.

Gagné, R. M., and L. J. Briggs. 1979. *Principles of Instructional Design*. 2nd ed. New York: Holt, Rinehart and Winston.

Toastmasters International. 2008. *From Speaker to Trainer: Coordinator's Guide*. Rancho Santa Margarita: Toastmasters International.

20 Adopting the New Paradigm: Virtual Training and M-learning

20.1 DEFINITION OF VIRTUAL TRAINING

Welcome to the world of virtual training, which is expanding every day!

Although the title of this chapter mentions the new paradigm, virtual training itself is nothing new. It is the many new forms of virtual training that are the new paradigm, especially m-learning, or mobile learning.

Traditionally, virtual training has been referred to as a training delivery method using a simulated virtual environment. By this definition, flight simulators are early examples of virtual training. They were developed at the beginning of the 20th century. At that time flight simulators consisted of mechanical equipment, levers, and controls. The training was virtual as it was simulated, as opposed to being carried out in a real airplane. Later in the century, computer technology was developed and flight simulators began to deliver a virtual reality experience, which is one form of virtual training as it is known nowadays.

Today, virtual training is delivered through a variety of methods and technologies. It can be defined as any training that takes place in a simulated or online environment. With regard to the interaction between trainer and trainee, virtual training may be trainer-led or self-directed. Delivery may be synchronous or asynchronous. The trainer and trainee may or may not be at the same physical location. This chapter will focus on distance learning where the trainer and trainee do not meet in person; they may meet in cyberspace.

20.2 EVOLUTION OF DISTANCE LEARNING

Distance learning was developed in higher education and spread into corporate training. For many years courses have been delivered asynchronously. Trainees obtain course materials by accessing recordings on tapes, CDs, DVDs, and computer servers. Asynchronous delivery is still widely used but as technology in videoconferencing, web conferencing, and live media streaming advances and costs decrease, synchronous delivery in whole or in part gains popularity. Meanwhile, as the delivery methods evolve, so does terminology. "Computer-based training" (CBT) has turned into "e-learning" to encompass a wider range of delivery modes by electronic means. It is defined as the delivery of information and instruction through the use of computer network technology over the intranet or Internet (Welsh et al., 2003, 246). Later on, with laptop computers having outsold desktop computers and the exponential growth of other mobile devices, m-learning has become the wave of today and the future.

20.3 CAVEATS IN IMPLEMENTING VIRTUAL TRAINING

Before deciding that a particular training course can be delivered virtually, you want to determine the suitability of using this delivery method by analyzing the advantages and disadvantages. Additionally, trainee assessment and course design are two areas that deserve special attention in most types of virtual training.

20.3.1 Suitability

The earlier statement that m-learning is the wave of the future does not mean that virtual training will be the exclusive delivery method in training. There will be in-person training for many years to come, and blended training is also an alternative. The fundamental principle that everything in a training plan should support the learning objectives always applies. Before deciding whether to use any form of virtual training, consider the pros and cons.

20.3.1.1 Advantages of Virtual Training

The most oft-cited advantage of virtual training is savings in cost and time. A large group of trainees can be trained at a relatively low cost since classroom facilities are not required and course materials are not printed. Thousands of geographically dispersed trainees can be accommodated simultaneously while interactivity and collaboration functions can be incorporated. The training can be brought to the trainee's location, reducing time and expense of travel by trainer and trainee. In most virtual training, expensive demo equipment is not needed or the cost is much less than real equipment. The practice environment is safe.

When virtual training is self-paced and self-directed, conflict in scheduling is not an issue, making this delivery method convenient for shift workers and telecommuters. It is suitable for younger employees that are especially adept in the digital world. Regardless of the time when individual trainees access a training course, standardized and consistent course materials are delivered, reinforced by links to searchable databases and resources for quick reference. On the administrative side, it is easy to track and document virtual training using a learning management system or other means of electronic record keeping.

20.3.1.2 Disadvantages of Virtual Training

Trainee's lack of access to hardware used to be an impediment to virtual training because in some organizations, not all employees are assigned individual computers. With the proliferation of smartphones, practically all employees would have devices that they can use to participate in virtual training. Reliable high-speed data network connection, bandwidth and processing power for live video streaming, and storage capacity of the device may remain challenging in some cases.

The initial costs of designing virtual training can be high. It is much more expensive in terms of the trainer's time than many incidental trainers would expect.

Converting existing course materials to formats for e-learning or m-learning may not be effective. Design, testing, and retesting before deployment to ensure that both content and user interface are satisfactory can consume an enormous amount of resources. When virtual training is trainer-led, the trainer must be skillful in this delivery method and for many trainers accustomed to the traditional classroom setting, practice and adaptation are vital. Some virtual training requires expensive hardware, such as the modern flight simulator.

User interface must meet regulatory standards. For example, nondiscrimination laws require features that enable persons with disabilities to use the programs. Usability studies must include special populations such as older workers to reduce resistance and encourage adoption. Content security and copyright issues must also be addressed. These technical aspects may call for consultation with an instructional technologist that has expertise in developing virtual training courses.

Virtual training is often self-directed. During course development, the trainer must decide when to allow learner control and when to prescribe program control in such things as how many practice examples trainees can work on. Lee and Lee (1991, 496–497) find that a learner-control strategy is superior for knowledge review, whereas a program-control strategy is preferred for knowledge acquisition. A high degree of learner control, as opposed to program control, can adversely affect learning effectiveness under certain conditions, for example, when trainees have poor metacognitive skills (Levinson et al., 2007, 499–500). Moreover, an individual's internal locus of control may affect satisfactory completion of a self-directed course (Parker, 1999). According to Frankola (2001), reasons for corporate e-learners to drop courses include insufficient motivation, lack of time, distractions at work, inability to access intranet from home, inadequate management oversight, poor course design and support, and inexperienced trainer. Course materials must be presented in ways that overcome distraction or possible feelings of isolation. Interactivity, milestones, and deadlines need to be incorporated to motivate trainees to complete the training in a timely manner. Finally, integrity of tests and assessments is hard to preserve in the virtual environment.

20.3.2 TRAINEE ASSESSMENT

Virtual training should meet the same standard as traditional training, which means that trainees should be tested for the knowledge and skills gained during this learning process. Advances in technology enable many types of testing to be performed virtually, from quizzes to games. You will need to select those most appropriate for the training topic and ensure, to the extent possible, that individuals taking a test in a course delivered asynchronously are the persons that they claim to be. Software attempts in various ways to identify test-takers but for the most part it is an honor system unless a proctored examination is administered.

Overall, technology facilitates assessment and feedback. As an example, software can be programmed to conduct an adaptive test in which a trainee receives a question based on whether the previous question is answered correctly. Another enhancement is the ability for trainees to receive model answers after a test. Thede et al. (1994, 302) have found that trainees appreciate a full explanation or demonstration of the correct steps in performing a task, rather than just a reference where they must locate and read later.

20.3.3 Course Design

Sitzmann et al. (2008, 289) point out that training is more effective when technology-delivered courses are useful and interesting to trainees. Mayer (2003, 310) recommends the following design principles for e-learning, based on results of many studies on the ability to apply what is learned after training:

- *Multimedia principle:* Use multimedia, for example, animation and narration, to explain concepts.
- *Modality principle:* Present animation and narration rather than animation and on-screen text.
- *Contiguity principle:* Animation and narration should be presented at the same time if the information elements must be integrated in working memory to be intelligible.
- *Personalization principle:* Use conversational style, not formal narration, in the spoken words.
- *Coherence principle:* Extraneous video and audio like background music or sounds, even if they are interesting, would have a negative effect.
- *Redundancy principle:* Adverse outcome would result from adding redundant text in multiple modes.
- *Pretraining principle:* If the training involves explanation of how something works, begin by describing the components of the object or phenomenon.
- *Signaling principle:* Use signaling in the narration that may include announcing headings of each section of the presentation in a deeper voice.
- *Pacing principle:* Incorporate a feature such as a button to let a trainee advance to the next section when he or she is ready, instead of automatic timing.

Design recommendations for mobile apps in m-learning are discussed later in this chapter.

20.4 STRATEGIES FOR THE VIRTUAL CLASSROOM

Although self-directed virtual training is convenient, there are circumstances when direct interaction between trainer and trainee is desired to enhance human interaction. Coaching and mentoring are examples. Videoconferencing has been available since the 1970s. With new technology, voice over Internet protocol (VoIP), web conferencing, and live video streaming have been developed to support computer and mobile platforms, enabling delivery of lectures, demonstrations, and discussions in virtual classrooms in much the same way as in a physical classroom. Proper preparation and presentation are needed even more for the virtual classroom than for the physical classroom because of a less controllable environment and potential technical glitches. It is not uncommon during web conferences for audio or video

Adopting the New Paradigm: Virtual Training and M-learning   185

problems to arise that degrade the quality of the training or result in inaccessibility by some trainees. Choose the service that offers the functionality most suited to your needs (Better Trainers Inc., 2013), then prepare meticulously for the virtual training event and deliver it with style.

20.4.1 Functionality

If your organization already has a service provider for web conferencing, you probably would be working with that system. If not, you must decide which features and capabilities will be appropriate for your training courses. The following are features commonly used but may not be offered by all service providers.

- *Chat room:* Any participant of the web conference can type and send a message to all participants or a specific person.
- *Whiteboard:* The trainer, or another person given control by the trainer, can draw or write on a virtual board in a similar fashion as a whiteboard in a traditional classroom.
- *Annotation tools:* The trainer, or another person given control by the trainer, can use a pen, highlighter, or arrow to draw attention to an area on the screen.
- *Polling:* The trainer can pose questions to the trainees, as in a survey, and display the results on the screen.
- *Raised hand:* A trainee can raise a virtual hand so the trainer or moderator knows that this trainee has a question or needs help.
- *Breakout rooms:* Trainees can hold subgroup discussions and collaborate on special assignments in virtual breakout rooms.
- *Granting control:* The trainer can give a trainee or the moderator control over a document displayed on the trainer's screen, using the mouse or other input device.
- *Changing presenter:* The trainer can grant a trainee or the moderator "presenter status" so everyone would be viewing that person's screen instead of the trainer's.
- *Muting:* The trainer can mute the microphones of trainees one at a time or all at once, which may be desirable during a lecture.
- *Removing trainee:* The trainer can disconnect a disruptive trainee and prevent the individual from reconnecting to the event.
- *Monitoring trainee engagement:* The trainer's console displays the percentage of trainees who focus on something other than the training, such as Internet surfing.
- *Recording and playback:* Trainees who were present at the virtual training event can review it again and others who were unable to attend can view the event.

Two advanced features may be useful for some training topics or industries. They are high-definition content display and live video streaming. High-definition content is video or image recorded at a high resolution. If a radiologist is training junior associates on the interpretation of x-ray images, he or she would want to use high-definition content display to preserve viewing quality. While many web conferencing services can display on the screen the faces of the trainer, moderator, and trainees who have activated their webcams, as of this writing only the premium services can stream a live video of the training event. This feature comes in handy if you want to enable trainees from remote locations to attend a training event virtually during the time when you conduct the event for a local group in a traditional classroom.

20.4.2 Preparation

Most of the preparations for the physical environment apply to the virtual environment, with modifications.

20.4.2.1 Advance Site Inspection

The "site" is the physical environment in which you deliver the training, as well as the virtual classroom.

Depending on the topic of the training, you may be facilitating the virtual training event from your office or another site. Wherever the physical site is, check it out in advance to be sure that it has the services and equipment you need, such as high-speed Internet access. A wired Internet connection and a USB headset are preferred to obtain the highest quality of audio and video transmission.

As for the virtual classroom, plan which software features you would use and perform a test meeting with an assistant (see "Other Logistics" below), a colleague, or yourself using a second computer or mobile device. Note that some services would allow a meeting participant to join the meeting using a mobile device but the meeting organizer's console is not supported for a mobile platform. If possible, record the test meeting so you can review it and make improvements as necessary.

20.4.2.2 Room Layout and Seating

If you will appear on video during the virtual training event, be aware of the surroundings that will show up on the screen of your trainees, such as clutter on your desk or on a bookcase behind you. If you will be seated, make sure that the seat is comfortable and adjustable to a suitable height.

20.4.2.3 Lighting, Noise, and Climatic Conditions

Lighting of your physical location is very important when it will be on camera. The lighting should be sufficient, which to a certain extent depends on the specifications and capabilities of the camera. There should be no glare—you will be amazed how glare that you do not notice in the surroundings will show up in a photo or video, so it is wise to look around through the lens of a camera rather than your naked eye. This is another reason why recording and reviewing your test meeting will be helpful.

With or without video, you need to eliminate distracting noise that may be audible to your trainees. Maintain comfortable climatic conditions for yourself so you can be relaxed while facilitating the virtual training event.

Adopting the New Paradigm: Virtual Training and M-learning

20.4.2.4 Other Logistics

You will find it extremely helpful to identify an assistant to serve as a moderator at your virtual training event. This person can prioritize questions received in the chat area and direct them to you, monitor trainee engagement, and handle technical issues if they occur. The virtual classroom would run more smoothly with such an arrangement. The assistant does not have to be at the same physical location as you.

Course materials can be uploaded to the web conferencing server for trainees to download or view before the virtual training event. As discussed in Chapter 6, there are pros and cons of distributing course materials prior to a training event. If the materials are made available, let the trainees know where and how they can access and download. Provide trainees with clear directions for entering the virtual classroom and eliminating distractions, encourage them to test their audio and video connection, as applicable, before the start time of the training event, and explain how they can get help in case of a technical problem.

20.4.3 Delivery

Most presentation and facilitation guidelines for in-person training are also applicable to virtual training. One example is starting and ending a training event on time. Trainees may be late because they have not followed your instruction to allow time for testing their audio and video connection. You should still start the virtual training event on time unless there is a major technical issue.

For course delivery in the video mode, a few additional guidelines should be noted (Simon, 2013, 7; Toastmasters International, 1992, 28–31):

- Wear clothing that contrasts with the background color of your physical environment so you are not blended in.
- Avoid wearing clothing or a tie that has stripes or plaids or jewelry that produces glare as they are distracting on the screen.
- Dress in a medium-colored suit and pastel shirt or blouse; avoid black, white, or bright red colors.
- Sit in an open chair or stand unless you are giving a demonstration; do not stand behind a desk.
- In a seated position, lean forward slightly; relax but do not slouch or swing back and forth in a swivel chair.
- When standing, keep your shoulders straight and distribute your weight evenly on both feet, except when your demonstration requires another posture.
- If you move around, stay within the range of the camera; some webcams can follow movements, others are fixed and have a narrow range.
- Ensure that any object you show is also visible within the range of the camera and, if possible, is in a color that stands out from the background and not causing glare.
- Project your voice with vigor so that your voice will not sound monotonous; check and set the correct volume on the microphone.

- Be careful that unwanted sound is not picked up by the microphone, for example, when you shuffle papers or accidentally hit the microphone.
- Refrain from gesturing directly and close to the camera as your hands and arms will be distorted.
- Smile whenever appropriate—you will appear relaxed and friendly!

In addition, it is good at the beginning of a virtual training event to remind trainees to "turn off" anything at their end that may generate distracting noise. Occasionally, trainees call in from home, and everyone hears barking dogs or crying babies! As a default setting, some software automatically mutes all participants except the organizer. During the training, remember that it is more difficult to capture and maintain audience attention in the virtual classroom than the traditional classroom, so be concise in word usage, whether spoken or written in the chat area.

20.5   GROWTH IN M-LEARNING

M-learning, or mobile learning, has been defined and interpreted in many ways. EDUCAUSE (2010, under "What is it?") states, "Mobile learning, or m-learning, can be any educational interaction delivered through mobile technology and accessed at a student's convenience from any location." As in other methods of training delivery, m-learning can be used for target audiences that include employees, customers, and prospects.

It is estimated that the worldwide mobile worker population will increase to 1.3 billion by 2015 (Crook et al., 2011). With the changing characteristics of the work force, the prominence of m-learning has increased rapidly. Its popularity is fueled, in no small part, by the widespread adoption of the use of mobile devices by consumers and businesses globally. For example, 60% of the American population in 2010 went online using a laptop or mobile phone (Pew Research Center, 2010). In Africa, the second largest mobile market in the world after Asia, the number of mobile connections has grown an average of 30% annually and m-learning initiatives have helped improve access to education where the number of schools is limited in remote areas (GSM Association, 2011). In the United States alone, the market for m-learning products and services reached almost $1 billion in 2010, with a five-year compound annual growth rate of 13.7% (Adkins, 2011).

The exponential growth of the market for mobile devices, especially smartphones, offers tremendous opportunities for the use of m-learning in training delivery. Mobile apps are particularly versatile and appropriate.

20.6   ADVANTAGES OF MOBILE APPS

Some of the advantages of using mobile apps in training are as follows:

- Ease of access
- Timeliness of information
- Engagement of trainees
- Support of training activities

- Chunking of content
- Availability of software

20.6.1 Ease of Access

The integration of mobile devices in people's lives means that a trainee can use mobile apps to access training modules or other information without special hardware or equipment. The nature of these devices allows easy access to training materials anytime, anywhere. Learning can take place whenever a trainee has a few minutes of "downtime," whether it is waiting for a plane to arrive or for a restaurant to prepare a food order.

20.6.2 Timeliness of Information

"Learning and performance support are morphing into a single application in a contemporary version of on-the-job training" (Adkins, 2011, 5). Tools available in m-learning facilitate such a convergence. Just-in-time information can be incorporated into mobile apps to improve communication and enhance trainee performance. For example, the safety data sheet for a hazardous substance can be obtained by a laboratory employee instantaneously on a smartphone at the time of cleaning up a chemical spill, without having to search for the information on a desktop computer or in a binder.

20.6.3 Engagement of Trainees

The fact that mobile devices have become almost omnipresent among the population regardless of geographic and cultural diversity promotes trainee engagement. A trainee is more likely to access the apps when a mobile device is at his or her fingertips than when the individual must retrieve information or training modules through a less portable device. There is also the suggestion that the novelty aspect may induce the "feeling of play" and stimulate trainee interest (Laborda and Royo, 2009, 143).

20.6.4 Support of Training Activities

A study by the Pew Research Center (2010) has found that as of May 2010, 34% of mobile phone owners in the United States play games using their phones. Games, simulation, and similar activities incorporated into m-learning apps would be a natural extension of activities with which trainees are already familiar. Collaborative learning is made easy when every member of a team has a smartphone that can be used for both activities and communications.

20.6.5 Chunking of Content

While more and more facts and data are readily retrievable in this information age, people's attention span is becoming shorter and shorter. This is particularly true in a generation that has grown up texting and tweeting. Research has demonstrated the benefits of matching training to adult attention span through chunking, for example, by delivering three 20-minute sessions instead of one 1-hour instruction (Murphy, 2008). Mobile apps are conducive to delivering materials in chunks. As a matter of fact, they are most suitable for engaging trainees for a brief time, perhaps only a few minutes, as in providing instant information on policies and procedures or extension of concepts and skills learned in an earlier course. For instance, a sales professional who has been trained in a new product line can obtain more advice on the best ways to present the new products while preparing for a meeting with a prospect.

20.6.6 AVAILABILITY OF SOFTWARE

In addition to special software developed for m-learning, there is a wealth of software applications that can be incorporated into m-learning apps. Examples are the e-book reader and geolocation software. Some software is proprietary and does not work across different mobile devices and platforms. The trend, however, is that more cross-platform programs are being developed.

20.7 EXPECTATIONS OF MOBILE APP FEATURES

Whether your goal is to select mobile apps from an outside source or to develop apps in-house, the characteristics of well-designed apps must be understood.

A few years ago, CBT was the buzzword. Now m-learning is the hot topic. CBT and m-learning share something in common that is unfortunate—both are "victims" of the myth that training materials used in these delivery methods can be produced simply by converting existing materials into a different format. For a long time, much of CBT has been based on PowerPoint presentations designed for in-person training. A similar approach is found where e-learning materials are converted to a "lite" version to facilitate retrieval on a mobile device, similar to websites that have a mobile version. To maximize the benefits of m-learning, content should be specifically designed for mobile delivery. This does not mean that one should discard the training materials from in-person and e-learning and start from scratch. It does mean that the presentation of the materials deserves special thought. The contents of existing materials can be leveraged in m-learning for cost-effectiveness (Poulos, 2013).

Why is m-learning not e-learning "lite"? One of the major differences between e-learning and m-learning is that the former is designer-centered, whereas the latter is trainee-centered. Oftentimes e-learning has complicated menu structures. Also, e-learning content is written on web pages for a target group of trainees. There is still a target audience in m-learning; however, the content should be customized for each individual in the group. This is because to take full advantage of the power of m-learning, a trainee should be able to access only the information on the part of the "page" that is relevant to his or her need at the moment (Poulos, 2013). Apps for m-learning should be designed to meet this and other trainee expectations. Although every individual is different, the following features are fundamental to gaining trainee acceptance:

- Personalization
- Multimedia
- Interactivity
- Integration
- Support

20.7.1 Personalization

Personalization encompasses customization in the areas of context, content, and navigation. It reduces information overload. It is particularly relevant when an on-the-job trainee encounters a situation where access to the right information in the right place at the right time is critical. The good news is that m-learning software enables you or your instructional technologist to easily assemble and reuse content and media assets in different ways to suit different trainees and situations (Poulos, 2013).

20.7.2 Multimedia

Trainees using mobile apps expect materials to be presented in multimedia. When used properly, rich content that includes text, graphics, audio, and video is conducive to learning. Simply adding an animated pedagogical agent can be effective in enhancing the learning process and improving learning outcome (Atkinson, 2002, 426). Trainees may also need content that can be downloaded for offline use. Learning management systems offer the convenience of placing content in repositories and adapting the media format for delivery to various mobile devices. It must be emphasized that when multiple modalities are used simultaneously, the different components should be interrelated. For instance, learning outcome will improve, when an image is presented along with text, only if the image is relevant to the text from the trainee's point of view (Dubois and Vial, 2000, 163); otherwise the image might create meanings that confuse the trainee in a self-directed learning environment.

20.7.3 Interactivity

The popularity of social media and virtual communities reflects users' affinity for sharing and collaboration. Mobile apps would be expected to integrate highly interactive functions. This is good for organizations as the likely results would be more efficient team cooperation and higher team and individual performance.

20.7.4 Integration

Since a trainee may switch from one mobile device to another or to a desktop computer, it should be possible to retrieve and reuse content on a different device or delivery platform once the content has been created, including trainee-generated content, such as notes and blog posts. Trainees would not want to use an app that does not have this flexibility.

20.7.5 Support

An advantage of delivering m-learning is that it has no boundary in time or place, so 24/7 access is the norm. Even though m-learning is self-paced for the most part, trainees may need to consult a subject-matter expert (SME) or technical support at some point. Availability of these services, also 24/7, is preferred especially around the time of initial launch. Technical problems cause trainee frustration (Welsh et al., 2003, 255).

20.8 DESIGN OF MOBILE APPS FOR TRAINING

Training apps on various topics for many professions are available in the market. As an incidental trainer, you may want to save development time and cost by purchasing existing apps. The content of any third-party product must be screened for accuracy. The functionalities of the apps should also be tested before they are distributed to your trainees.

Whether you purchase apps or develop them in-house, the apps should aim at meeting trainee expectations as described above. Unfortunately, until the frontier of mobile technology pushes out further, there are technical constraints. The following are factors to be considered in designing mobile apps that trainees would want to use:

- Trainee experience
- Screen size
- Connection speed
- Storage capacity
- File format
- Font style

20.8.1 Trainee Experience

Despite its many advantages, much of m-learning is self-directed and requires trainees' self-discipline, motivation, and effort. The training is delivered in surroundings that have more distractions than the traditional classroom. The appearance of the mobile app's landing page as well as other content must have visual appeal to grab and maintain attention. A "search" function is essential to assist self-directed trainees in finding what they need. A "help" function can also be embedded in the app. Such user-friendly features can reduce incidents when SME or technical support is needed.

20.8.2 Screen Size

Although mobile devices vary in types and sizes, the most ubiquitous is still the smartphone. A mobile app's page layout and menu system should be simple so that elements are not cluttered on a small screen and navigation using small buttons or a touch screen is easy. Use short titles to avoid wrapping and concise text to reduce scrolling. Consistency in the organization of content and interface would help trainees navigate through the screens to explore a topic in-depth or move to another topic (Advanced Distributed Learning Initiative, 2013).

20.8.3 Connection Speed

Another reason for simplicity of page layout is that the connection speeds of mobile data networks are usually slower than, say, the speed of a cable network connected to a desktop computer. Wi-Fi networks sometimes used by mobile devices vary in speed. Taking too long to download content may drive a user away.

20.8.4 Storage Capacity

Simple page design and small file size also reduce the demand for storage space. In spite of the rapid advances in hardware for mobile technology, storage space is limited compared with a desktop computer. Content to be downloaded and stored in a mobile device, so that it can be referenced later offline, should occupy a small storage space. This aspect may not be an issue for the average users of mobile devices as they are usually connected to the data networks. It may be important for trainees in some occupations working at locations that may not have network access.

20.8.5 File Format

Not all file formats are supported by all mobile devices. Also, some file formats are only viewable on certain devices using additional software. To ensure maximum compatibility, Wilson et al. (2011) suggest staying with the most common file formats, such as pdf, doc, docx, ppt, jpg, gif, png, mp3, and mp4.

20.8.6 Font Style

Since mobile devices generally do not have many fonts installed, it is advisable to use a basic font. Arial is preferable to Times New Roman since a sans serif font is easier to read, especially on a small screen. Avoid using symbols.

20.9 INCLUSION OF PERFORMANCE SUPPORT TOOLS

The types of digital content that can be incorporated into apps for m-learning are numerous. External links to additional resources can be included as well. The following are a few examples of content applicable to just-in-time and on-the-job training:

- Decision support tools
- Regulatory and consensus standards
- Policies and procedures
- Technical documentation
- Operating instructions
- Alerts, reminders, and checklists

Along with reference materials, pertinent photos and videos can be included for "live" demonstration. Mobile augmented reality technology using geotagging to overlay contextual content on physical locations is valuable, for example, in emergency management to show evacuation routes.

Content inserted into a workflow process improves efficiency and facilitates teamwork. A mobile app designed to train auditors, for instance, may include a feature whereby with the touch of a button, a new auditor can be connected to the lead auditor to request guidance on specific issues. When used for this purpose, the content should be organized in a manner parallel to the actual workflow process.

20.10 SUCCESSFUL DEPLOYMENT OF VIRTUAL TRAINING

Before full deployment the pilot course should be tested with a sample group of trainees. Obtain trainee feedback and implement changes as appropriate. Ensure that support is available to trainees from the pilot phase to full deployment and thereafter.

Innovative Learning Group Inc. (2011) recommends a few extra steps to ensure successful deployment of mobile apps:

- Publicize the app on the Internet or intranet and emphasize its benefits to the trainees.
- Assure trainees that they can access the "help" function within the app if necessary.
- Post a demonstration or practice simulation online for the trainees to see and try out the app's features and functions.
- Compile an FAQ based on feedback from the sample in the pilot course and early adopters among the target group; post it along with tips for using the app.

20.11 SUMMARY

The successful use of any form of virtual training, as other delivery methods, requires that the method supports the learning objectives which, in turn, are aligned with the goals of a comprehensive training program and the mission of the organization. Lewis and Orton (2000, 49) point out that in adopting innovations, organizations should evaluate the innovation's relative advantage, compatibility, simplicity, trialability, and observability. These factors are favorable for the adoption of virtual training. With new technologies that enable interaction and collaboration in cyberspace, training can be conducted effectively in the virtual environment.

The enormous opportunities offered by virtual training cannot be condensed into one chapter, nor does technology stand still. After you have developed and delivered a well-designed, trainee-centered course, continuous monitoring, evaluation, and refinement are imperative in a world of ever-changing technological advances.

ACKNOWLEDGMENT

Part of this chapter is adapted from Wan (2013), a proceedings paper of Safety 2013, a Professional Development Conference of the American Society of Safety Engineers.

REFERENCES

Adkins, S. S. 2011. "The US Market for Mobile Learning Products and Services: 2010–2015." Ambient Insight LLC. http://www.ambientinsight.com/Resources/Documents/Ambient-Insight-2010-2015-US-Mobile-Learning-Market-Executive-Overview.pdf.

Advanced Distributed Learning Initiative. 2013. *Mobile Learning Handbook.* Accessed March 11. https://sites.google.com/a/adlnet.gov/mobile-learning-guide/home.

Atkinson, R. K. 2002. "Optimizing learning from examples using animated pedagogical agents." *Journal of Educational Psychology* 94 (2): 416–427. doi:10.1037//0022-0663.94.2.416.

Better Trainers Inc. 2013. "Choosing a Virtual Classroom Environment." Accessed March 23. http://www.bettertrainers.org/resources/choosing-a-virtual-classroom-environment/.

Crook, S. K., J. Jaffe, R. Boggs, and S. D. Drake. 2011. "Worldwide Mobile Worker Population 2011-2015 Forecast." Abstract. International Data Corporation (IDC). http://www.idc.com/getdoc.jsp?containerId = 232073#.UUgzdzeO71s.

Dubois, M., and I. Vial. 2000. "Multimedia design: The effects of relating multimodal information." *Journal of Computer Assisted Learning* 16 (2): 157–165.

EDUCAUSE. 2010. "7 Things You Should Know about Mobile Apps for Learning." http://www.educause.edu/ir/library/pdf/ELI7060.pdf.

Frankola, K. 2001. "Why online learners drop out." *Workforce* 80 (10): 53–60.

GSM Association. 2011. "African Mobile Observatory 2011: Driving Economic and Social Development through Mobile Services." http://www.gsma.com/publicpolicy/wp-content/uploads/2012/04/africamobileobservatory2011-1.pdf.

Innovative Learning Group Inc. 2011. "Delivery Method Selection Guidelines." http://www.innovativelg.com/content/secure/viewpdf.aspx?f = ILG_Delivery_Method_Selection_Guidelines_White_Paper.pdf.

Laborda, J. G., and T. Magal Royo. 2009. "Training senior teachers in compulsory computer based language tests." *Procedia Social and Behavioral Sciences* 1 (1): 141–144. doi:10.1016/j.sbspro.2009.01.026.

Lee, S.-S., and Y. H. K. Lee. 1991. "Effects of learner-control versus program-control strategies on computer-aided learning of chemistry problems: For acquisition or review?" *Journal of Educational Psychology* 83 (4): 491–498. doi:10.1037/0022-0663.83.4.491.

Levinson, A. J., B. Weaver, S. Garside, H. McGinn, and G. R. Norman. 2007. "Virtual reality and brain anatomy: A randomised trial of e-learning instructional designs." *Medical Education* 41 (5): 495–501. doi:10.1111/j.1365-2929.2006.02694.x.

Lewis, N. J., and P. Orton. 2000. "The five attributes of innovative e-learning." *Training & Development* 54 (6): 47–51.

Mayer, R. E. 2003. "Elements of a science of e-learning." *Journal of Educational Computing Research* 29 (3): 297–313.

Murphy, M. 2008. "Matching workplace training to adult attention span to improve learner reaction, learning score, and retention." *Journal of Instruction Delivery Systems* 22 (2): 6–13.

Parker, A. 1999. "A study of variables that predict dropout from distance education." *International Journal of Educational Technology* 1 (2). http://education.illinois.edu/ijet/v1n2/parker/index.html.

Pew Research Center. 2010. "Mobile Access 2010." http://pewinternet.org/Reports/2010/Mobile-Access-2010.aspx.

Poulos, D. 2013. "The Top 10 Reasons your Mobile Learning Strategy Will Fail." Xyleme Inc. Accessed March 13. http://www.xyleme.com/blog/the-top-ten-reasons-your-mobile-learning-strategy-will-fail.

Simon, C. 2013. "Are You Memorable?" White Paper. Rexi Media. Accessed January 29. http://www.reximedia.com/Portals/67770/docs/whitepaper.pdf.

Sitzmann, T., W. J. Casper, K. G. Brown, K. Ely, and R. D. Zimmerman. 2008. "A review and meta-analysis of the nomological network of trainee reactions." *Journal of Applied Psychology* 93 (2): 280–295. doi:10.1037/0021-9010.93.2.280.

Thede, L. Q., S. Taft, and H. Coeling. 1994. "Computer-assisted instruction: A learner's viewpoint." *Journal of Nursing Education* 33 (7): 299–305.

Toastmasters International. 1992. *Communicating on Television*. Rancho Santa Margarita: Toastmasters International.

Wan, M. 2013. "Successful use of mobile apps in m-learning and risk communication." In *Safety 2013, Proceedings of the 2013 ASSE Professional Development Conference*. Des Plaines: American Society of Safety Engineers. CD-ROM.

Welsh, E. T., C. R. Wanberg, K. G. Brown, and M. J. Simmering. 2003. "E-learning: Emerging uses, empirical results and future directions." *International Journal of Training and Development* 7 (4): 245–258.

Wilson, E., L. Day, L. Hives, J. Kelleher, and R. Lilleker. 2011. "Best Practices for Mobile-Friendly Courses." Blackboard Inc. http://www.blackboard.com/getdoc/59e6f603-876e-4833-9757-d22c6bffd092/Best-Practices-for-Mobile-Friendly-Courses.aspx.

Epilogue

The original plan was to provide a list of resources in a final chapter for your future reference. The rate at which new information becomes available and previous information changes is incredible. For this reason, no resource list is compiled other than the reference list at the end of each chapter. Instead, you are invited to visit my personal website http://www.MargaretWan.com where sample forms and resources are posted.

Practice makes perfect. Incidental trainers may not use many of the training skills as often as professional trainers. Every training event, however, is a great opportunity for you to practice training like a pro!

Appendix A: Training Needs Analysis Sample Form

Appendix A: Training Needs Analysis Sample Form

Training Needs Analysis

Location		☐ Indoor		☐ Outdoor
Department				
Interviewee		Job title		
Facility tour	Item	Yes	No	N/A
	Workstation design, equipment, and supplies appropriate	☐	☐	☐
	Illumination adequate	☐	☐	☐
	Background noise controlled	☐	☐	☐
	Temperature and humidity within comfort range for most people	☐	☐	☐
	Work area free from recognized hazards (physical, chemical, biological, or ergonomics-related)	☐	☐	☐
	Notes:			
Regulatory and policy requirements				
Job requirements ☐ Job analysis attached	Knowledge:			
	Skills:			
	Abilities:			
	Other characteristics:			
Previous training				
Gap analysis	What *do* employees know?			
	What *should* employees know?			
	How *do* employees perform?			
	How *should* employees perform?			
	Why is there a discrepancy?			
Training needed	Job Title		Topic	
Other recommendations				
Trainer's name/title				
Trainer's signature		Date		

Reproduced by permission from *Fundamentals of Training: Design, Development, Delivery.*
© 2010 Better Trainers Inc. and Margaret Wan.

Appendix B: Task Analysis Sample Form

Task Analysis

Location			Dept	
Training needs analysis	Report date		Report #	
Task analyzed/goal			Performer	
			Job title	
Document review	Item		Yes	No	N/A
	Job description		☐	☐	☐
	Standard operating procedures Reference number(s):		☐	☐	☐
	Equipment user guide(s) Name, make, and model:		☐	☐	☐
Prerequisites	Skills:
	Supplies:
	Tools:
Task inventory: preparation phase	1.
	2.
	Describe any deviations from standard or recommended procedures.
Task inventory: performance phase	1.
	2.
	3.
	4.
	Describe any deviations from standard or recommended procedures.
Task inventory: follow-up phase	1.
	2.
	Describe any deviations from standard or recommended procedures.
Task inventory validated	☐ No ☐ Yes, by_____(name and title)
Environmental constraints	
	Describe any attempts to resolve constraints and the outcomes.
Trainer's name/title	
Trainer's signature		Date	

Reproduced by permission from *Fundamentals of Training: Design, Development, Delivery.*
© 2010 Better Trainers Inc. and Margaret Wan.

Appendix C: Training Plan Sample Form

Training Plan
(Page 1)

Course title						
	☐ New		☐ Revised		☐ Agenda attached	
Training frequency	☐ One-time		☐ Annual		☐ Other _____	
Task or course synopsis						
	Goal or application					
	Task analysis		☐ N/A		☐ Attached	
Target audience	Job titles/functions					
	Prerequisites					
	KSAOs					
	Shift work		☐ No		☐ Yes	
	Number of years in job	< 1	_____%	6–10	_____%	
		1–5	_____%	> 10	_____%	
	Education	HS/GED	_____%	Graduate	_____%	
		Associate	_____%	Profession	_____%	
		Bachelor's	_____%	Other	_____%	
	Gender	Male	_____%	Female	_____%	
	Generation	Millennials	_____%	Boomers	_____%	
		Gen Xers	_____%	Silents	_____%	
	Non-English speaker %					
	Special accommodation					
	Additional notes					
Learning or performance objectives	After completion of this course, participants should be able to: 1. 2. 3. 4.					
Successful course completion criteria				Continuing ed credit (if applicable)		
Instructional strategies	Lecture/ panel	Group discussion	Demo/ practice	Role-playing	Self-guided discovery	Collab learning
Objective 1	☐	☐	☐	☐	☐	☐
Objective 2	☐	☐	☐	☐	☐	☐
Objective 3	☐	☐	☐	☐	☐	☐
Objective 4	☐	☐	☐	☐	☐	☐
	Notes on the use of the selected instructional strategies:					

Reproduced by permission from *Fundamentals of Training: Design, Development, Delivery.*
© 2010 Better Trainers Inc. and Margaret Wan.

Appendix C: Training Plan Sample Form

Training Plan
(Page 2)

Training aids and media	Type	Description/title	
	Handout	☐ Electronic	☐ Hard copy
	Slides		
	Video/audio DVD/CD		
	Easel pad/white board (+ markers)		
	Model/prop		
	Other aids and media		
Other equipment	☐ Data projector	☐ Screen	☐ Easel
	☐ Computer (trainer)	☐ Speakers	☐ DVD/CD player
	☐ Computer (trainee)	☐ Internet access	☐ Other_____
Physical environment	☐ Indoor	☐ Virtual-synchronous	☐ Blended
	☐ Outdoor	☐ Virtual-asynchronous	
	Room layout	☐ Theater	☐ H-square
		☐ Rounds	☐ U-shape
		☐ Classroom	☐ Conference
		☐ Other_____	☐ Not applicable
	Special requirement		
Testing methods	Pretest ☐ Written ☐ Oral		
	Posttest ☐ Written ☐ Oral		
	Skills validation		
	Other		
Course evaluation	By participants	☐ Written	☐ Oral
	By _____	☐ Written	☐ Oral
Trainer's observations and suggested changes	Learning objectives		
	Instructional strategies		
	Training materials		
	Testing methods		
	Delivery style		
	Overall evaluation		
Other comments			
Trainer's name/title			
Trainer's signature		**Date**	

Reproduced by permission from *Fundamentals of Training: Design, Development, Delivery.*
© 2010 Better Trainers Inc. and Margaret Wan.

Appendix D: Course Evaluation Sample Form

Course Evaluation

Your opinion is important for continuous improvement of the course. Please print all write-in answers.

Course title:					
Date:					
Trainer:					
Participant (optional):					
Please check the box that best describes how you feel about the statements below.	Strongly agree	Somewhat agree	Neutral or N/A	Somewhat disagree	Strongly disagree
The course is relevant to your professional work.	☐	☐	☐	☐	☐
The course met the learning objectives stated.	☐	☐	☐	☐	☐
The presentation was organized in a logical flow.	☐	☐	☐	☐	☐
The course materials stimulated learning.	☐	☐	☐	☐	☐
The exercises or activities were effective.	☐	☐	☐	☐	☐
The right amount of time was given each topic.	☐	☐	☐	☐	☐
The trainer was knowledgeable of the subject.	☐	☐	☐	☐	☐
The trainer communicated ideas clearly.	☐	☐	☐	☐	☐
The trainer motivated audience participation.	☐	☐	☐	☐	☐
The trainer maintained audience interest.	☐	☐	☐	☐	☐
The trainer responded to questions effectively.	☐	☐	☐	☐	☐
The trainer started and ended on time.	☐	☐	☐	☐	☐
What did you learn from this course that is the most useful?					
What topic, if any, should be added to or eliminated from this course?					
How can this training course or the trainer be more effective?					
Additional comments:					

Thank you for your feedback!

Reproduced by permission from *Fundamentals of Training: Design, Development, Delivery.* © 2010 Better Trainers Inc. and Margaret Wan.

Index

A

ABCD formula, 31–32
   audience, 31
   behavior, 31
   condition, 31–32
   degree, 32
   examples, 32
Adaptation, 30
Advance site inspection, 186
Affective domain, 26–28
   organization, 28
   receiving, 27
   responding, 28
   value, characterizing by, 28
   valuing, 28
Aids to training, 21, 47–57
   audio presentations, 50–51
   computers, 53–54
   costumes, 52
   games, 53
   handouts, 47–48
   Internet, 53–54
   media presentations, 54–56
   models, 52
   props, 52
   slide presentations, 48–50
   video, 50–51
Alternative formats, training plan, 21
Analysis, 26
Analysis of needs, 8–12
   audience, 12
   facility tour, 9
   goals of, 8–9
   internal policies, 10
   interviewing personnel, 9–10
   job analysis, reviewing, 10
   performance gap, 11–12
   regulatory requirements, 10
   remedial actions, 11–12
   surveys, 9–10
   topics, 12
   training record, reviewing, 10
Analysis of task, 13–17
   goals of, 13
   intellectual tasks, 16–17
   job description, distinguished, 13
   method, 13–14
   multiple task analyses, 17
   performance, 14–16
   equipment, 15
   inventory, 16
   job description, 14–15
   listing steps, 15–16
   observation, 15
   performance, 16
   standard operating procedures, 15
Application, 26
Apps, mobile, 192–193
   connection speed, 193
   expectations of, 190–192
   file format, 193
   font style, 193
   integration, 191
   interactivity, 191
   multimedia, 191
   personalization, 191
   screen size, 192
   storage capacity, 193
   support, 192
   trainee experience, 192
Arrival time, 174–175
Assessment, 67–75
   approaches to testing, 67–70
   criterion-referenced testing, 68–69
   criticality of, 67
   formative testing, 69–70
   norm-referenced testing, 68–69
   objectivity, 74
   posttest, 68
   pretest, 68
   reliability, 70–71
   subjectivity, 74
   summative testing, 69–70
   testing methods, 71–74
      essay, 72–73
      fill-in-the-blank, 72
      judgment testing, 73
      multiple choice, 72
      observation, 74
      oral explanation, 72–73
      report, 74
      role-playing, 73–74
      short answer, 72
      simulation, 73
   validity, 70–71
Attendance, 173
Audience, 12, 31, 83
Audio presentations, 50–51
Auditor's evaluation, 89

B

Baby boomers, learning style, 137–140
Behavior management, 85
Benefits of training, 4–5, 109–110
    productivity, 109
    quality, 109–110
    safety, 110
Body language, 81–82
Breakout session, 170–171
Business case presentation, 115

C

Climatic conditions, 186
CLT. *See* Cognitive load theory
Cognitive domain, 24–26
    analysis, 26
    application, 26
    comprehension, 25
    evaluation, 26
    knowledge, 24–25
    synthesis, 26
Cognitive load theory, 154–155
Collaborative learning, 41–42
Communication skills, 157–158
Communication strategies, 150–151
Compensation, 108
Comprehension, 25
Computers, 53–54
Condition, 31–32
Cone of experience, 35–36
Conformity, 146
Constructivism, 155–156
Continuing education credit, 21
Continuous quality improvement, 102–103
Controlled experiment, 100–101
Cost-benefit/benefit-cost ratio, 110
Costs of materials, equipment, 108
Costs of training, 108–109
Costumes, 52
Course completion criteria, 21
Course design, 184
Course evaluation, 21, 87–94, 99
    assessment, 94
    auditor's evaluation, 89
    criticism, asking for, 87
    data relevancy, 93–94
    designing surveys, 90–93
    evaluations, 87–89
    response rate, 92–93
    sample form, 207–208
    supervisor's evaluation, 89
    timing, 93
    trainee's evaluation, 89
    trainer's self-evaluation, 88

CQI. *See* Continuous quality improvement
Credits, continuing education, 21
Criterion-referenced testing, 68–69
Criticism, asking for, 87
Cultural intelligence, 148
Cut scores, 122–123

D

Data relevancy, 93–94
Debate, 168
Defects in training, 3
Degree, 32
Demonstration, 40
Designing surveys, 90–93
Directing event, 173–179
    arrival time, 174–175
    attendance, 173
    ending on time, 178
    events of instruction, 176
    greeting trainees, 175
    guest speakers, 178
    interactive methods, 176–177
    logistics, 174
    preparation, 173
    starting on time, 175
    visual aids, 177–178
Disruptive trainees, 159–160
Distance learning, 181–182
Diversity, cultural, 145–151
Documentation, 101–102

E

Educational materials, 163–169
Emotions, 83
Ending on time, 178
Entertaining materials, 163–169
Enthusiasm, 169–171
Environment, 59–66
    advance site inspection, 60–61
    climatic conditions, 62–63
    control, 60–64
    equipment, 64
    evacuation routes, 63–64
    lack of success, 59
    lighting, 62
    logistics, 64
    noise, 62
    room layout, 61–62
    safety, 63–64
    sanitation, 63
    seating, 61–62
    supplies, 64
    water, 63

Index

Equipment, 15
Essay, 121
Evaluation, 26
Evaluation of course, 21, 87–94
   assessment, 94
   auditor's evaluation, 89
   criticism, asking for, 87
   data relevancy, 93–94
   designing surveys, 90–93
   evaluations, 87–89
   response rate, 92–93
   supervisor's evaluation, 89
   timing, 93
   trainee's evaluation, 89
   trainer's self-evaluation, 88
Event, directing, 173–179
Events of instruction, 176
Expenses, 109

F

Facilitation, 77–86
   audience, 83
   behavior management, 85
   body language, 81–82
   emotions, 83
   feedback, 85–86
   knowledge, 83
   organization, 79–80
   preparation, 78
   purpose, 82–83
   questioning, 84
   responding to questions, 84–85
   style, 79–82
   visuals, 82
   vocabulary, 80
   vocal variety, 80–81
Facilitator, 153–162
Facility tour, 9
Facility usage, 108–109
Feedback, 85–86, 160
Fill-in-the-blank, 121
Fixed-choice questions, 118–120
   matching, 120
   multiple choice, 119
   true/false questions, 120
Font style, mobile apps, 193
Formative testing, 69–70
Forms
   course evaluation, 207–208
   task analysis, 201–202
   training needs analysis, 199–208
   training plan, 203–205
Four E's, 163–172
   breakout session, 170–171
   debate, 168
   educational materials, 163–169
   entertaining materials, 163–169
   enthusiasm, 169–171
   game, 168–169
   general discussion, 170
   icebreaker, 163–164
   multimedia presentation, 164–165
   problem-solving activity, 167–168
   role-playing, 166
   self-guided discovery, 165–166
Frontline supervisors, 114–115
Functionality, 185–186

G

Games, 53, 168–169
Gender roles, 146
General discussion, 170
Generation Xers, learning style, 138, 140
Generational learning, 137–144
   baby boomers, 137–140
   generation Xers, 138, 140
   generations, 138–143
   millennials, 138, 140–141
   silents, 137–139
Generations
   learning style, 141–143
   learning styles, 138–143
Goals of needs analysis, 8–9
Goals of task analysis, 13
Governing board, 113–114
Greeting trainees, 175
Group discussion, 39–40
Guest speakers, 178

H

Handouts, 47–48

I

Icebreakers, 163–164
Imitation, 30
Incidental trainers, 4
Instructional design, 157
Instructional strategies, 21, 35–45
   collaborative learning, 41–42
   cone of experience, 35–36
   defining, 35
   demonstration, 40
   group discussion, 39–40
   learning objectives, 42
   learning styles, 36–37
   lecture, 38–39
   on-the-job training, 38
   panel, 38–39

practice, 40
role-playing, 40–41
self-guided discovery, 41
situational constraints, 43
target audience, 42–43
trainer's skills, 43
ubiquitous lecture, 35
Integrated training program, 95
Intellectual tasks, 16–17
Interactive methods, 176–177
Interactivity, mobile apps, 191
Internal policies, 10
Internal rate of return, 113
Internet, 53–54
Interviewing personnel, 9–10
Inventory, 16
IRR. *See* Internal rate of return

J

Job analysis, reviewing, 10
Job description, 14–15
distinguished, 13

K

Keys to successful training, 5–6
Knowledge, 24–25, 83

L

Lack of control, 65–66
distractions, 65
furniture design, 65–66
Learning objectives, 20, 23–34, 42
ABCD formula, 31–32
audience, 31
behavior, 31
condition, 31–32
degree, 32
examples, 32
affective domain, 26–28
Bloom's taxonomy, 24–26
analysis, 26
application, 26
comprehension, 25
evaluation, 26
knowledge, 24–25
synthesis, 26
cognitive domain, 24–26
domains of learning, 24–30
educational objectives, 24–30
Krathwohl's taxonomy, 26–28
organization, 28
receiving, 27
responding, 28
value, characterizing by, 28
valuing, 28
learning objectives, 30–34
psychomotor domain, 28–30
adaptation, 30
imitation, 30
observation, 29
practice, 30
SMART principle, 32–34
actionable, 33
measurable, 33
results-oriented, 33
specific, 33
trainee-centered, 33–34
training plan, 23
writing, 23
Learning styles, 36–37, 137–144
Learning theories, 154–156
Lecture, 38–39
Lighting, 186
Listening, 158
Listing steps, 15–16
Logistics, 174, 187

M

M-learning, 188–190. *See also* Mobile apps
access, 189
activities, support for, 189
content chunking, 189–190
engagement, 189
software, 190
timeliness, 189
Matching, 120
Media presentations, 54–56
Metric validity, surveys, 128
Microphones, 174
Middle managers, 114–115
Millennials, learning style, 138, 140–141
Mobile apps, 192–193
connection speed, 193
expectations of, 190–192
file format, 193
font style, 193
integration, 191
interactivity, 191
multimedia, 191
personalization, 191
screen size, 192
storage capacity, 193
support, 192
trainee experience, 192
Models, 52
Multicultural work force, 145–152
communication strategies, 150–151
comprehension, 151

Index

conformity, 146
cultural diversity, 145–147
cultural intelligence, 148
gender roles, 146
nonverbal techniques, 149–150
power distance, 146–147
trainees, 147–148
translations, 149
uncertainty, acceptance of, 146
Multimedia, mobile apps, 191
Multimedia presentation, 164–165
Multiple choice, 119
Multiple task analyses, 17

N

Needs analysis, 8–12
   audience, 12
   facility tour, 9
   goals of, 8–9
   internal policies, 10
   interviewing personnel, 9–10
   job analysis, reviewing, 10
   performance gap, 11–12
   regulatory requirements, 10
   remedial actions, 11–12
   surveys, 9–10
   topics, 12
   training record, reviewing, 10
Needs assessment, 7–18
Net present value, 112
Network usage, 108–109
Noise, 186
Nonverbal techniques, 149–150
Norm-referenced testing, 68–69
NPV. *See* Net present value

O

Observations, 15, 21, 29, 100
On-the-job training, 38
Open-ended questions, 120–121
   essay, 121
   fill-in-the-blank, 121
   oral explanation, 121
   short answer, 121
Oral explanation, 121
Organization, 28, 79–80
Organizational impact, 103
Organizational support, 107–116
   benefits of training, 109–110
      productivity, 109
      quality, 109–110
      safety, 110
   business case presentation, 115
   compensation, 108
   cost-benefit/benefit-cost ratio, 110
   costs of materials, equipment, 108
   costs of training, 108–109
   employees, 115
   expenses, 109
   facility usage, 108–109
   frontline supervisors, 114–115
   governing board, 113–114
   middle managers, 114–115
   network usage, 108–109
   senior management, 113–114
   value, 107–108

P

Panel, 38–39
Parts of learning plan, 20–21
Payback period, 111
Performance, 16
Performance assessments, 122
Performance gap, 11–12
Performance objectives, 20, 23–34
   ABCD formula, 31–32
      audience, 31
      behavior, 31
      condition, 31–32
      degree, 32
      examples, 32
   affective domain, 26–28
   Bloom's taxonomy, 24–26
      analysis, 26
      application, 26
      comprehension, 25
      evaluation, 26
      knowledge, 24–25
      synthesis, 26
   cognitive domain, 24–26
   domains of learning, 24–30
   educational objectives, 24–30
   Krathwohl's taxonomy, 26–28
      organization, 28
      receiving, 27
      responding, 28
      value, characterizing by, 28
      valuing, 28
   learning objectives, 30–34
   psychomotor domain, 28–30
      adaptation, 30
      imitation, 30
      observation, 29
      practice, 30
   SMART principle, 32–34
      actionable, 33
      measurable, 33
      results-oriented, 33
      specific, 33
      trainee-centered, 33–34
   training plan, 23

Performance support tools, 193–194
Physical environment, 21, 59–66
    advance site inspection, 60–61
    climatic conditions, 62–63
    control, 60–64
    equipment, 64
    evacuation routes, 63–64
    lack of success, 59
    lighting, 62
    logistics, 64
    noise, 62
    room layout, 61–62
    safety, 63–64
    sanitation, 63
    seating, 61–62
    supplies, 64
    water, 63
Pilot test, 133
Plan, 19–21
    alternative formats, 21
    continuing education credit, 21
    course completion criteria, 21
    course evaluation, 21
    general information, 20
    instructional strategies, 21
    learning objectives, 20
    parts, 20–21
    performance objectives, 20
    physical environment, 21
    structure, 19–21
    target audience, 20
    testing methods, 21
    trainer's observations, 21
    training aids, 21
Posttest, 68
Power distance, 146–147
Practice, 30, 40
Preparation, 78, 173, 186–187
Presentation, 77–86
    audience, 83
    behavior management, 85
    body language, 81–82
    emotions, 83
    facilitation, 83–86
    feedback, 85–86
    knowledge, 83
    organization, 79–80
    preparation, 78
    purpose, 82–83
    questioning, 84
    responding to questions, 84–85
    style, 79–82
    visuals, 82
    vocabulary, 80
    vocal variety, 80–81
Pretest, 68
Problem-solving activity, 167–168

Productivity, 109
Program validation, 95–104
    controlled experiment, 100–101
    course evaluation, 99
    criteria, 96–98
    documentation, 101–102
    integrated training program, 95
    observation, 100
    quasi-experiment, 100–101
    reaction survey, 99–100
    test results, 99
    tools, 98–101
Psychomotor domain, 28–30
    adaptation, 30
    imitation, 30
    observation, 29
    practice, 30
Purpose, 82–83

Q

Quality improvement, 102–103
Questioning, 84
Questioning techniques, 158–159
Questions, responding to, 84–85

R

Reaction survey, 99–100
Receiving, 27
Reflection, 66
Regulatory requirements, 10
Reliability testing, 117–124
    case study, 121–122
    cut score, 122–123
    fixed-choice questions, 118–120
        matching, 120
        multiple choice, 119
        true/false questions, 120
    open-ended questions, 120–121
        essay, 121
        fill-in-the-blank, 121
        oral explanation, 121
        short answer, 121
    performance assessment, 122
    situational judgment, 121–122
    weight of evidence, 117
Remedial actions, 11–12
Responding to questions, 28, 84–85
Role-playing, 40–41, 166
Room layout, 186

S

Sample forms
    course evaluation, 207–208
    task analysis, 201–202

Index

training needs analysis, 199–208
training plan, 203–205
Screen size, mobile apps, 192
Seating, 186
Self-guided discovery, 41, 165–166
Senior management, 113–114
Short answer, 121
Silents, learning style, 137–139
Site inspection, 186
Situational constraints, 43, 56
Situational judgments, 121–122
Slide presentations, 48–50, 177
SMART principle, 32–34
    actionable, 33
    measurable, 33
    results-oriented, 33
    specific, 33
    trainee-centered, 33–34
Software, M-learning, 190
Standard operating procedures, 15
Starting on time, 175
Storage capacity, mobile apps, 193
Structure, training plan, 19–21
Style, 79–82
Subjectivity, 74
Successful training, keys to, 5–6
Suitability, 182–183
Summative testing, 69–70
Supervisor's evaluation, 89
Support, organizational, 107–116
    benefits of training, 109–110
        productivity, 109
        quality, 109–110
        safety, 110
    business case presentation, 115
    compensation, 108
    cost-benefit/benefit-cost ratio, 110
    costs of materials, equipment, 108
    costs of training, 108–109
    employees, 115
    expenses, 109
    facility usage, 108–109
    frontline supervisors, 114–115
    governing board, 113–114
    middle managers, 114–115
    network usage, 108–109
    senior management, 113–114
    value, 107–108
Survey completion, 132
Survey delivery, 131–132
Surveys, 9–10, 125–135
    data quality, 125–132
    design, 125
    designing, 90–93
    metric validity, 128
    mode, 126–128
    pilot test, 133
    question content, 128–129
    question presentation, 129–131
    response rate, 131–132
Synthesis, 26

T

Target audience, 20, 42–43, 55
Task analysis, 13–17
    goals of, 13
    intellectual tasks, 16–17
    job description, distinguished, 13
    manufacturer's instructions, 15
    method, 13–14
    multiple task analyses, 17
    performance, 14–16
        equipment, 15
        inventory, 16
        job description, 14–15
        listing steps, 15–16
        observation, 15
        performance, 16
        standard operating procedures, 15
    sample form, 201–202
Test results, 99
Testing, 67–75
    approaches to testing, 67–70
    criterion-referenced testing, 68–69
    criticality of, 67
    formative testing, 69–70
    norm-referenced testing, 68–69
    objectivity, 74
    posttest, 68
    pretest, 68
    reliability, 70–71
    subjectivity, 74
    summative testing, 69–70
    testing methods, 71–74
        essay, 72–73
        fill-in-the-blank, 72
        judgment testing, 73
        multiple choice, 72
        observation, 74
        oral explanation, 72–73
        report, 74
        role-playing, 73–74
        short answer, 72
        simulation, 73
    validity, 70–71
Testing methods, 21, 71–74
    essay, 72–73
    fill-in-the-blank, 72
    judgment testing, 73
    matching, 72
    multiple choice, 72
    multiple select, 72
    observation, 74

oral explanation, 72–73
ordering, 72
performance, 73
report, 74
role-playing, 73–74
short answer, 72
simulation, 73
true false, 72
Tools for program validation, 98–101
Topics, 12
Trainee achievement, 153
Trainee assessment, 183–184
Trainees, 147–148
Trainee's evaluation, 89
Trainer's observations, 21
Trainer's self-evaluation, 88
Trainer's skills, 43, 55–56
Training aids, 21
Training needs analysis, sample form, 199–208
Training plan, 19–21, 23
    ABCD formula, 31–32
        audience, 31
        behavior, 31
        condition, 31–32
        degree, 32
        examples, 32
    affective domain, 26–28
    alternative formats, 21
    Bloom's taxonomy, 24–26
        analysis, 26
        application, 26
        comprehension, 25
        evaluation, 26
        knowledge, 24–25
        synthesis, 26
    cognitive domain, 24–26
    continuing education credit, 21
    course completion criteria, 21
    course evaluation, 21
    educational objectives, 24–30
    general information, 20
    instructional strategies, 21
    Krathwohl's taxonomy, 26–28
        organization, 28
        receiving, 27
        responding, 28
        value, characterizing by, 28
        valuing, 28
    learning objectives, 20, 30–34
    parts, 20–21
    performance objectives, 20
    physical environment, 21
    psychomotor domain, 28–30
        adaptation, 30
        imitation, 30
        observation, 29
        practice, 30
        sample form, 203–205
    SMART principle, 32–34
        actionable, 33
        measurable, 33
        results-oriented, 33
        specific, 33
        trainee-centered, 33–34
    structure, 19–21
    target audience, 20
    testing methods, 21
    trainer's observations, 21
    training aids, 21
Training record, reviewing, 10
Translations, 149
True/false questions, 120
Two-way communication, 157–158

U

Ubiquitous lecture, 35
Uncertainty, acceptance of, 146

V

Validation, program, tools, 98–101
Validity, 70–71
Validity testing, 117–124
    case study, 121–122
    cut score, 122–123
    fixed-choice questions, 118–120
        matching, 120
        multiple choice, 119
        true/false questions, 120
    open-ended questions, 120–121
        essay, 121
        fill-in-the-blank, 121
        oral explanation, 121
        short answer, 121
    performance assessment, 122
    situational judgment, 121–122
    weight of evidence, 117
Value, characterizing by, 28
Value of training, 107–108
Valuing, 28
Video, 50–51
Virtual training, 181–196
    course design, 184
    defining, 181
    delivery, 187–188
    distance learning, 181–182
    functionality, 185–186
    preparation, 186–187
    suitability, 182–183
    trainee assessment, 183–184

Index

Visual aids, 177–178
Visuals, 82
Vocabulary, 80
Vocal variety, 80–81

W

Wireless microphones, 174
Word usage, 92
Writing learning objectives, 23

X

Xers. *See* Generation Xers

Y

Younger employees, 138, 143, 182

Z

Zero net present value, 113